Newnes Telecommunications Pocket Book

Newnes
Telecommunications
Pocket Book

Third edition

Steve Winder, BA, MSc, CEng, MIEE, MIEEE

Newnes

OXFORD   AUCKLAND   BOSTON
JOHANNESBURG   MELBOURNE   NEW DELHI

Newnes
An imprint of Butterworth-Heinemann
Linacre House, Jordan Hill, Oxford OX2 8DP
225 Wildwood Avenue, Woburn, MA 01801-2041
A division of Reed Educational and Professional Publishing Ltd

A member of the Reed Elsevier plc group

First published 1992
Reprinted 1995, 1996
Second edition 1998
Third edition 2001

British Library Cataloguing in Publication Data

A catalogue record for this book is available from the British Library

ISBN 0 7506 52985

FOR EVERY TITLE THAT WE PUBLISH, BUTTERWORTH-HEINEMANN
WILL PAY FOR BTCV TO PLANT AND CARE FOR A TREE.

Typeset by Laser Words, Madras, India
Printed in Great Britain

Contents

commonly a plastic polymer film able to retain at least part of an electric charge placed on it.

The carbon microphone depends on the variation in resistance of a quantity of carbon granules, confined in a small chamber and subjected to pressure variations via a membrane which in return has a sound pressure wave impressed on it. This microphone has been used for many years in telephone systems.

The moving-coil microphone is similar in principle to the loudspeaker (see Section 2.3) and is used in situations requiring high-quality sound reproduction. Variable capacitance (condenser) microphones in which a metal diaphragm under the influence of sound pressure formed part of a DC polarized capacitor have largely been replaced by the electret.

The electret [2] is typically polarized by a voltage derived from the exchange line. Included in the microphone package is an FET connected as a source follower to buffer the microphone from the capacitance of its connection cable and input of the microphone amplifier. The sensitivity is comparable with that of a dynamic (i.e. moving-coil) microphone and is typically -60 dB at 1 kHz where 0 db $\equiv$ 1 V/μ bar.

Microphones may be ominidirectional or designed for a specific field sensitivity pattern. The utility of a directional pattern in some applications (e.g. radio sports commentators) is obvious.

The electret principle has also been employed for hydrophones and pressure switches.

2.5 Earpiece

Current through the energizing coil produces a magnetic flux in the iron yoke. One end of the yoke is polarized magnetic North, while the other end is polarized South. The magnetic flux from the permanent magnet flows through the armature and causes it to be attracted to the North end of the iron yoke, and repelled from the South end. As the current through the coil alternates, the speaker diaphragm moves to and fro.

The telephone earpiece is commonly a moving-iron device, an elementary form of which is illustrated in *Figure 2.2*.

2.6 Payphones

The environment of payphones varies considerably from the high-revenue public call office (PCO) to the low-utilization rural PCO and

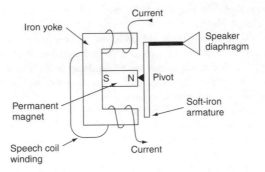

Current

Iron yoke

Speaker
diaphragm

S N Pivot

Permanent
magnet

Soft-iron
armature

Speech coil
winding

Current

Figure 2.2

the payphone installed in a customer's premises for the use of visitors
or clients. Different models to cope with the requirements differ in
their robustness of construction and protection against vandalism, and
in their sophistication, rather than in any basic operational sense.

Technically the improvements in payphones have been concentrated on their main weaknesses: the inflexibility of the coin handling,
their inability to handle trunk and overseas calls without operator
intervention, and mechanical unreliability. Again the improvements
have been made possible by the availability of integrated circuits and
microprocessors.

The first area of improvement is in the coin checking, where electrical eddy current and electrostatic checks have replaced mechanical
gauging as a means of coin validation. This has not only improved
reliability and reduced the risk of coin jamming, but enabled a range
of coinage to be fed into the coin slots. The coins are totalled and
held pending the successful connection of the call.

On less sophisticated equipment the customer must insert an
amount to cover the expected cost of the call, and the coins must
be released into the collection box before conversation can start,
that is the system is incapable of returning change at the completion
of the call. Call charging to a telephone customer is achieved by
metering charge units, each charge unit being indicated by an impulse.
The interval between impulses is therefore the time allowed for one
unit of charge, and varies according to the call destination, being
relatively long on local calls and much shorter on international calls,
also varying between peak, standard and cheap tariff periods. Sending
back signals derived from the metering pulses to the payphone
therefore enables the cost of the call to be debited against the
cash inserted, without operator intervention. The more sophisticated

equipment will indicate the cash balance remaining, enabling the caller to feed in further cash if required without interrupting the call. At the end of the call this gives the caller the option of recovering the balance remaining, or making further calls to use the balance.

One of the main problems with payphones, particularly on high-revenue sites, is full cash boxes. The more complex equipment provides automatic alert signals to a central control as this condition is approached, and is also able to signal fault conditions. The alternative approach is to eliminate cash collection and use prepaid phone cards, and payphones operating on this principle are in widespread use. Payphones using credit card payment are popular for overseas calls where the cost is generally higher.

With normal billing it is more convenient to change the tariff by altering the cost per unit than to change the time per unit, leading to costs per unit involving fractional coin values. This can be a problem with payphones, requiring software in the payphone which requires updating if the tariff rate is changed. This situation is simplified if the software is concentrated at the exchange rather than in the payphone, and a trend in this direction is likely as the number of digital exchanges increases.

2.7 Cordless telephones CT1, CT2

The cordless telephone consists of a base unit which provides the normal interface for connection to the two-wire PSTN. A rest is provided on the base unit for the handset, but the only electrical connection between base unit and handset are two contacts to provide charging for the handset batteries from the base unit power supply. Speech and signalling are achieved by radio links between base unit and handset, which can therefore be removed from the rest and used to make or receive calls anywhere within a specified radius of the base unit. To make this possible, the keypad and tone caller must be incorporated in the handset.

In addition to the microphone, earphone and keypad the handset has to contain a transmitter and receiver, signal processing, and batteries to provide the required power. This limits the transmitted power, and hence the operational range, which is of the order of 100 to 300 metres, according to local conditions. This is normally adequate for domestic use, and the restricted range reduces the risk of interference between users, and enables installations to operate without licensing.

14

A mandatory requirement for cordless telephones is a built-in security code, which is set by the manufacturer on a non-erasable memory. A handshake operation between base and handset prevents any possibility of the base unit being utilized by another handset for either incoming or outgoing calls.

The first-generation cordless phones, classified as CT1, are analogue instruments, using different frequencies in the base to handset direction, compared to the handset to base direction, thus providing duplex operation with different frequencies for the two directions. Since the power required exceeds that obtainable from line, most base stations are mains powered, with backup battery supply and charging facilities for rechargeable batteries in the handset. CT1 was intended to fulfil the need for a portable extension phone on a domestic installation. It has little attraction for the business or commercial user, and is not harmonized to any standard.

The next generation of cordless phones, classified as CT2, is a digital system capable of providing far more facilities than CT1, and designed to appeal to business users. In addition to its use as a portable extension with a base unit (as with CT1), CT2 offers the following facilities:

(a) Use as a personal portable extension telephone on a PBX, with a range, depending on local conditions, of around 200 metres.
(b) The fact that the transmitted signal is digital, using a baseband 64 kbit/s modulating signal, provides the basic facility for sending digital data. Initially this is possible only when used as an extension to a PBX or LAN with digital access.

The radio spectrum allocated for CT2 is 864–868 MHz, providing 40 100 kHz channels. Two-level FSK modulation is used with Gaussian filtering to reduce interchannel interface. CAI specifies a shift of between −14.4 and −25.2 kHz for logic 0, and +14.4 to +25.2 kHz for logic 1, with a bit rate of 64 kbit/s.

Speech is encoded using adaptive differential pulse code modulation (ADPCM) which digitizes the speech to a 32 kbit/s signal (Chapter 14). The sampling rate is 8 kHz, and 16 successive samples are assembled as a package, that is 64 bits representing 2 ms of speech. The package is then transmitted in a 1 ms burst, alternate milliseconds being used for transmissions in the two directions. The actual transmission rate is therefore 64 kbit/s, but by alternating transmissions in the two directions, full duplex operation is achieved using only the single FM channel.

CT2 was developed in the UK, but is now in use worldwide. The ETSI standard for a Pan-European scheme DECT (Digital Enhanced Cordless Telephone), uses the frequency band 1880–1900 MHz. DECT, however, has more in common, both in facilities provided and techniques used, with GSM than with CT1 and CT2, and is therefore dealt with in Chapter 4.

2.8 Answering machines

Answering machines are tape recorders which can record messages when a telephone is unattended. When called, the recorder completes the answering loop and sends back a pre-recorded announcement message, asking the caller to leave a message. At the conclusion of the announcement the answering machine provides a 'speak now' signal, and switches to the record tape, remaining in this condition until the caller clears.

Answering machines are available as 'standalone' units which can be added to a standard telephone installation, but are also available as composite units; one self-contained unit may include answering machine, telephone, keypad dialling, number storage and timing facility.

Some units provide an interrogation facility enabling a caller to request a playback of the recorded messages, subject to a security signal to prevent unauthorized access.

Some answering machines use magnetic tape for recording messages. Solid state memory is now used too.

Telecommunication service providers provide a 'call minder' service, whereby the telephone network answers a call and digitally stores messages on behalf of customers. This service saves the customer from buying an answering machine and can be set up to record calls at certain times (e.g. at night).

3 Telex

Telex provides a means of sending text messages between two locations. The Telex network is separate from the PSTN and as such is more resistant to computer hacking. The teleprinters at both locations simultaneously print out the message as it is sent, marking the beginning and end of the message with the identities of sending and receiving machines. This constitutes proof of delivery and has legal status. Alternative means of text message transmission such as E-mail or facsimile are not so robust. For this reason Telex remains in use by organizations that require legal status to their documents such as shipping companies and banks.

Conversation mode between Telex machines is also possible, so questions and answers are interleaved on the printed page. The page is continuous, rather than being cut into A4 or foolscap sized pieces, so no part of the message can be lost. This continuous page reinforces the legal status, since it is not possible to substitute any pages out of a long message. Dot matrix printers are used to ensure that the message is 'original'. A laser printer cannot be used for legal purposes, since the output from it is essentially the same as from a photocopier, and it would be impossible to guarantee that the original message was not modified and then photocopied. Using a dot matrix printer also allows carbon copies to be made which is useful if different departments need to see the message.

Nowadays, computers have the ability to send Telex messages. Dedicated Telex machines are still used but these are rarely the slow electro-mechanical type. Most machines have a VDU display and allow the user to edit a message using word processor software. This allows the message to be checked for accuracy before being sent. Received messages can be stored and forwarded without having to be retyped.

3.1 Telex operation

Telex enables alphanumeric characters to be transmitted and repro-duced as typed characters at the receiving terminal. The system has evolved from the mechanical teleprinter, where the send keys pro-duced the pulse sequence, and the pulse current directly operated the type heads at the receiving terminal. Here, ±80 volts DC was used as the line signal, and earth return working was possible enabling one line of a pair to be used for each direction, and duplex operation achieved.

The speed was therefore limited to that of the operator and the printer. The 5 unit International Alphabet No. 2 (IA2) was adopted (see Chapter 26), a space being indicated by +80 volts and a mark by −80 volts. The frame representing one character was increased to 7.5 units by a +80 volts start unit and a −80 volt stop signal, the latter occupying 1.5 units, as shown in *Figure 3.1(a)*. Each unit of the code is 20 ms, one character therefore requiring 150 ms and providing a speed of $6\frac{2}{3}$ characters per second and a transmission rate of 50 baud (50 off 20 ms units per second). The speed of 50 baud enabled the motors of the transmitter and receiver to be synchronized by the mains supply, although the system is inherently asynchronous, each character frame containing its own start and stop indicators.

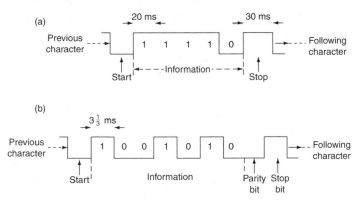

Figure 3.1   Telex Signals: (a) International Alphabet No. 2, letter 'k', speed 50 baud; (b) International Alphabet No. 5, letter 'k', speed 300 baud

The dependence on DC pulses required a separate switching network capable of handling the pulses, resulting in separate telephone and telex networks. This now has the advantage of providing a dedicated network, serving only telex, and enabling machines to be permanently connected and able to receive messages at any time. The low speed is still adequate for many commercial and industrial users, who require to send relatively short messages dealing with orders, transactions, etc.

The development of high-speed printers and the means of storing received signals has enabled the speed of the telex network to be extended to 300 baud. The increased speed can be utilized to provide operation with the 7 bit international alphabet No. 5 (Chapter 26).

A further development has been the replacement of ±80 volts signalling, by the adoption of SCVF (Single-Channel Voice Frequency). This uses speech band FSK with frequency division for the two directions to give duplex on the single local pair, and is to ITU-T Recommendation R20 (CEPT T/CD1–9).

3.2 Telex via the PSTN, private lines and radio

The original telex network, operating at 50 baud, limited character printing to 6.67 characters per second, compatible with the mechanical teleprinter. The development of low-cost information storage and high-speed printers has enabled very much faster telex machines to be manufactured. Extensions of the telex network to accept 300 baud transmissions can still be met with relatively simple telex terminals, requiring a print speed of 30 characters per second when using the 7 bit 1A5 code, which is well within the capabilities of a dot matrix printer.

Much faster speeds are, however, achievable but cannot be provided on the dedicated telex network. Higher-speed operation can be obtained by utilizing an entire voice-frequency channel, either over the PSTN or over a private line. Since this involves transmission over the frequency band 300–3400 Hz, a modem is required. An extensive selection of modems is available, providing a range of bit speeds and alternative modulation systems, as described in Chapter 6.

The speed in bits per second that can be transmitted within the confines of the voice-frequency channel depends on the complexity of the modulating system and on the quality of the transmission channel. It can also be restricted by the need to avoid certain frequencies used for signalling purposes. For this reason better results can usually be obtained on private lines than on a PSTN telephone line even though the actual private lines are PSTN lines rented to the user for exclusive use.

Simple FSK is satisfactory at bit rates of 300 per second, and in favourable conditions 600 bit/s is possible.

FFSK (Fast Frequency Shift Keying) employing frequencies of 1800 Hz and 1200 Hz to represent 0 and 1 must synchronize the changeovers to occur as the waveform passes through zero, that is 1 bit must be 1.5 cycles of 1800 Hz or 1 cycle of 1200 Hz, giving a bit rate of 1200 bit/s. This will provide transmission over most voice-frequency channels, with a character rate of 120 per second. This exceeds the speed of high-speed printers, and storage of the received characters prior to printing is necessary. Of the three phase modulation alternatives PSK (Phase Shift Keying), DPSK (Differential Phase Shift Keying) and QPSK (Quaternary Phase Shift Keying), the most important is QPSK. Phase shifts of 45°, 135°, 225° or 315° are applied every 1.5 cycles to an 1800 Hz carrier, and the four alternatives represent 4 dibits or 2 bits. The system therefore provides 2 bits for each phase shift interval, that is 2400 bit/s. A modem providing QPSK modulation can therefore provide a bit rate of 2400 per second on a voice-frequency circuit, but only if the circuit is of good quality and free from signalling frequencies within the 300 to 3400 Hz band. Using 1A5 code with start, stop and parity bits, this represents 240 characters per second.

Modern telex machines are normally designed to provide operation at 300 baud with 1A5 code, even if higher-speed capability for use with modems is provided, in order that they can be used on the telex network, or as a fallback if satisfactory operation at higher speed cannot be achieved. High-speed printing is provided by a dot matrix printer (using either 7×5 or 9×7 matrix), or a daisy wheel printer. Both require control circuits, the dot matrix to provide the correct firing pulses for the needle solenoids, and the daisy wheel to locate the wheel for printing. Speeds of the order of 50 and 30 characters per second are obtainable with the dot matrix and daisy wheel printers respectively. Storage is provided, which can be used either to store information prior to sending to line or to store information received prior to printing. A display panel (usually LCD) enables typed-in messages to be checked prior to transmission. The 1A5 code contains a range of control codes (Chapter 26), which can be transmitted by additional control keys on the keyboard. Some telex machines can accept high-speed data from a computer into their store, and act as an interface between computer and line.

Automatic facilities are normally provided for dialling, redialling if the required number is engaged, and sending at programmed times. Provided transmission speed is set at 300 baud, these sophisticated machines and their facilities may be used on the telex network, employing SCVF over the local line. When used on a private line or over the PSTN, duplex operation or semi-duplex can be worked, depending on the facilities provided by the modem. Some modems provide duplex with reduced speed on the return path where high speed in both directions is not required. Interface standards with modems are defined in EIA Recommendations RS232.

Slow speed teletype machines, like those used for Telex, remain in use but primarily in radio systems. The HF band radio is used for long range communication, the radio signals. Because of its popularity, each frequency band allocation is very restricted, hence single sideband speech, Morse code and teletype are used. Morse code and teletype provide a more robust service, compared with single sideband speech, because the HF band is prone to atmospheric disturbances that can distort speech and make it unintelligible.

4 Mobile phones and pagers

Mobile phones have become very popular, partly due to the decreasing size of the handset. Their popularity is also partly due to the improved quality of transmission and the provision of message storage facilities. A network of low power radio transmitters are arranged so that each one covers a small area, or cell. As users move towards the working range limit of a cell, the network switches the handset to operate on a frequency of a neighbouring transmitter. The network also switches the call in progress so that it reaches the same transmitter. Thus as the user moves from one area to another the call is not interrupted. Because the radio transmitters are low power, the same frequencies can be reused by distant transmitters without interfering with each other.

There are two basic technologies used in mobile phones: analogue and digital. The analogue system assigns a transmitter frequency to each telephone call. Thus, in a cell, there may be 20 or so frequencies in use. The two popular analogue systems are the Advanced Mobile Phone System, or AMPS, which was developed in the USA, and the Total Access Communication System, or TACS, which was developed in Europe. The TACS system is very similar to AMPS except that the system has more intelligence built in. The problem with analogue systems is that they are not secure – anyone with an FM receiver capable of tuning up to 1 GHz could listen in on conversations. Even worse, eavesdroppers could find out the user identity and use it to clone the telephone. Calls made on a cloned telephone are then billed to the unlucky original subscriber.

Although both the AMPS and TACS systems are compatible, they operate at different frequencies. This incompatibility has been one of the drivers for ensuring that future systems will have a global standard. Agreeing which standard to use is made difficult by national interests.

The digital system uses either time division multiplexing or spread spectrum techniques, so that a number of telephone calls share the same transmitter frequency. The Global System for Mobile Radio, or GSM, uses Time Division Multiple Access (TDMA). GSM was originally called Group System Mobile, after the name of a standards committee group in CEPT. The name change was intended to indicate the global use of such a system. GSM is only available in limited areas of the USA, in the guise of PCS-1900. A similar system to GSM is the Japanese Personal Digital Cellular (PDC). Another system that uses TDMA is IS-136, which is provided by AT&T Wireless and Southwestern Bell.

The phones supported by the AT&T system have the option of reverting to AMPS if a digital base station is not within range.

Competing with GSM, IS-136 and other TDMA systems is the newer Code Division Multiple Access (CDMA) system developed by Qualcomm in the USA, where it is known as IS-95. It is in use in the USA and parts of Asia, and has been trialled in Germany. An alternative CDMA protocol has already been adopted in Japan. CDMA is based on the spread spectrum technology that has been used by the military for many years. By using a wide bandwidth, and spreading the data across the whole band, the system is more immune to noise and less prone to fraud by cloning. It is likely that CDMA, whether IS-95 or not, will become the worldwide standard or Universal Mobile Telephone System (UMTS).

Locating GSM telephones to within 50 m is possible. A system called 'Cursor' has been developed at Cambridge University in the UK to perform this function. This can be considered a low-cost global positioning system (GPS). The FCC in the US is asking cellular service providers to give location information, so that calls to the emergency services can be traced to a block or neighbourhood. If a caller cannot be heard, at least they can be found.

4.1 Analogue cellular systems

Analogue cellular systems are now largely obsolete and replaced by digital systems. The requirement is for a system providing full duplex operation and the ability for a mobile to call, or be called by, any other mobile or subscriber connected to the PSTN. A number of analogue systems, differing in technical detail, are in operation. The system in the UK uses the TACS (Total Access Communications System) structure. It is an enhanced version of the American AMPS (Advanced Mobile Phone System) structure.

Ideally the operators would provide radio base stations covering the whole country, with overlapping coverage to ensure that calls could be made from any location. The first priority, however, is to cover the motorways, main trunk roads and urban areas from which the vast majority of calls originate. Such a network of overlapping service areas is referred to as a 'cellular network'.

4.1.1 TACS

The frequency bands allocated to TACS cellular are:

Mobile transmit   890–915 MHz ⎫ two-frequency duplex
Base transmit   935–960 MHz ⎭

Extended TACS, or ETACS covers a wider band of frequencies, due to the demand for additional channels. The additional frequencies are:

Mobile transmit	872–888 MHz
Base transmit	917–933 MHz

The voice-channel modulation is FM, maximum deviation 9.5 kHz.

Channel spacing is 25 kHz providing 1640 duplex channels. A small proportion of these channels is permanently allocated as signalling channels. Base transmit signalling channel is the forward signalling channel (FOCC) and the mobile transmit signalling channel is the reverse signalling channel (RECC). The signalling protocol is considerably more complex than for trunked PMR, requiring a higher bit rate of 8 kbit/s. This rate is too high for satisfactory FFSK and the basic data stream is Manchester encoded and transmitted using PSK modulation. In addition to the FOCC and RECC, some signalling during a call is carried out on the allocated voice channel, the channels being designated forward voice channel (FVC) and reverse voice channel (RVC). To combat adverse conditions messages are constructed with parity bits using BCH coding and messages are repeated as shown in *Table 4.1*.

Table 4.1 Composition of signalling frames

Signalling channel	Data bits	Parity bits	Repeats
FOCC	28	12	5
RECC	36	12	5
FVC	28	12	11
RVC	36	12	5

Cellular networks are made up from cell clusters. Different channels are allocated to each cell in a cluster, and each cell has its own FOCC and RECC. The cell patterns have to interlock to cover wide areas without gaps, a condition that is only satisfied by certain cluster patterns, such as those shown in *Figure 4.1*. The cell size, siting, aerial power and channel allocations all affect the co-channel interference and call for careful system design. Groups of cells are connected to a mobile switching centre from which the mobiles are controlled, and each mobile switching centre provides the line access to the PSTN and also is linked to all other mobile switching centres as shown in *Figure 4.2*.

The system operates as follows:

(a) *Mobile on idle*: Each mobile is programmed with the frequencies of all the FOCC, and when switched on the mobile scans these

24

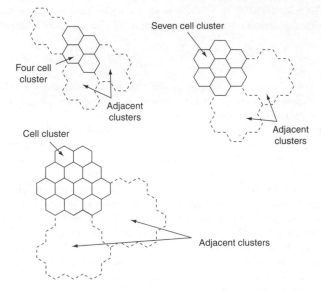

Figure 4.1 Cellular clusters. Different frequencies must be used for all channels in a cluster

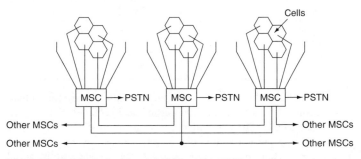

Figure 4.2 Cellular radio network: MSC, mobile switching centre

frequencies and selects the strongest signal, to which it synchro-nizes and locks. The mobile's identity code is transmitted over the RECC and its call location registered to enable incoming calls to be directed to the correct cell. The mobile then remains on standby, monitoring the incoming bit stream on the FOCC. If the received signal ceases to be satisfactory, as indicated by an unacceptable error rate or drop in level, the mobile repeats this procedure and transfers to another cell.

(b) *Incoming call*: The call is routed to the nearest mobile switching centre, and the cell in which the mobile is currently registered and the adjacent cells page the mobile using the paging sector of the FOCC. The mobile responds via the RECC, and when the call is ready to proceed a speech channel is allocated and the mobile instructed to switch to this channel for the duration of the call.

(c) *Outgoing call to a PSTN terminal*: The number required is keyed into the mobile's memory, or selected from a number store. The mobile seizes a free access channel. The channel numbers of the access channels are contained in the paging section of the FOCC, and a channel has first to be selected by scanning, and the information on access contained in the signal extracted. Seizure occurs when the channel busy/idle signal indicates that the access channel is free, and the call information is transmitted to the base station. The mobile transmitter is turned off but the mobile remains on the access channel until the call is set up, when the mobile is instructed to go to an allocated speech channel and start transmitting. Call routing is from the mobile switching centre into the PSTN, where it is routed to the required PSTN terminal. The interface between the mobile switching centre and the PSTN converts to the signalling system required by the PSTN exchange used for access.

(d) *Outgoing call to another mobile*: The procedure is as for (c) as far as the mobile switching centre. If the cellular network has sufficient voice-channel links between its switching centres the call can be routed direct to the switching centre in whose area the called mobile is registered. The called mobile is then paged and the call set up as in (b).

(e) *Mobile transmit power*: The received signal strength from a transmitting mobile is monitored by the base station. To reduce co-channel interference the base station can instruct the mobile to decrease or increase transmitted power.

(f) *Hand-off*: If the received signal strength falls below a satisfactory level the mobile is probably reaching the edge of the base station service area. The adjacent base stations are then instructed to monitor the signal strength of the channel, and a decision taken on transfer to an adjacent station. When transfer takes place the mobile is instructed to move to the new channel, and the location register is updated to show the new location of the mobile. This process is known as 'hand-off', and may be repeated several times if a mobile is moving along a boundary between two cells. Hand-off causes an interruption to the call in the order of 0.5 seconds.

Each TACS channel number, whether for speech or access channels, represents both go and return carriers for full duplex operation. The frequencies are paired with a separation of 45 MHz enabling channel switching to be achieved by a single change in synthesizer frequency. Modulation of the speech channel is analogue FM with a peak deviation of 9.5 kHz. This is high for a channel separation of 25 kHz, and improves the co-channel interference at the expense of adjacent channel interference. With careful choice of channel frequency allocation between cells the higher value of deviation has been found to provide the best overall performance.

4.1.2 AMPS

AMPS was developed by AT&T. It uses the same basic technology as the TACS system just described. There are a few subtle differences: each channel requires 30 kHz bandwidth, instead of the 25 kHz required for TACS. For a spectrum allocation of 20 MHz, AMPS provides 666 channels.

The transmit frequencies are different:

Base station	870–890 MHz
Mobile	825–845 MHz

FM deviation is ±12 kHz instead of the ±9.5 kHz used in TACS.

The control channel uses Manchester encoded FSK, with a peak deviation of ±8 kHz compared to ±6.4 kHz in the TACS system.

4.1.3 N-AMPS

Narrow-band–Advanced Mobile Phone Systems (N-AMPS) is based on the TACS and AMPS systems, as described in Sections 4.1.1 and 4.1.2, but uses a lower bandwidth. Motorola developed the N-AMPS system as an interim technology between analogue and digital. It has three times more capacity than the standard AMPS system and operates in the 800 MHz range, although largely superseded by digital systems.

4.2 Digital cellular systems – GSM

Analogue cellular systems have evolved with a common objective but sufficient differences in technical detail to render international operation difficult. To avoid this situation with the next-generation systems, at least as far as Europe is concerned, CEPT in 1982 set up a special committee 'Group Special Mobile' (GSM) to define and produce specifications for an internationally acceptable system. The new system was intended to take full advantage of the rapid developments occurring in the PSTN, that is the introduction of ISDN and digital exchanges, and to extend the advantages of digital communications to handle data as well as speech. The new system, known as GSM, is therefore digital for both traffic and signalling channels, whereas TACS is analogue for traffic and digital for signalling.

Note that GSM has been re-defined to 'Global System for Mobile', to take into account its global use.

The basic system parameters are

Base transmit frequency band	935–960 MHz	} two-frequency
Mobile transmit frequency band	890–915 MHz	} duplex

Carrier spacing	200 kHz
Channel multiplexing	eight per carrier TDM
Mobile power	20 mW to 300 W
Modulation	Minimum shift keying with Gaussian pulse (GMSK)

The frequency bands are the same as for TACS, and introduction in the UK therefore involves a progressive replacement programme, and not the setting up of a parallel operation.

The replacement of analogue by digital transmission on the traffic channel accounts for the major differences between TACS and GSM. The 200 kHz carrier spacing provides a total of only 125 alternative carrier frequencies, but the subdivision of each bit stream into eight TDM time slots, provides eight channels per carrier, thus equalling the total channel capacity of the TACS system.

Each time slot accommodates a transmit burst of 156.25 bits, occupying 540 microseconds, and repeated 216.68 times per second. Eight slots, representing the eight channels, make up a frame, and hence each frame is repeated 216.68 times per second.

This provides:

A bit speed during bursts of 289 350 bit/s
A total number of bits in the bit stream of 270 850 each second
An interval between bursts of 37 μs

The bandwidth required for an FM transmission is given in Chapter 26 as $2(f_m + f_D)$, where f_m and f_D are the modulating and deviation frequency respectively. The bandwidth is required to accommodate sidebands removed from the carrier by $f, 2f, 3f, \ldots$ and the formula derives from the fact that these assume appreciable amplitude at m values exceeding $0, 1, 2, \ldots$ respectively as indicated by the Bessel functions $J_1(m), J_2(m), J_3(m), \ldots$. The bandwidth actually increases in steps and the formula is therefore accurate for high values of m, and applies to analogue speech modulation where the lower frequencies have values of m exceeding 1, and hence a number of sidebands. However, with single-frequency modulation at low modulation index ($m < 1$) the formula ceases to be accurate, the $2f_m$ term being negligible up to an m value of about 0.5 (see Chapter 26). The bandwidth is therefore effectively $2f_m$ and not $2(f_m + f_D)$ for values of m up to 0.5; this is exploited in GMSK.

Shaping the rectangular waveform of the bit stream to Gaussian pulses reduces the harmonics, and compresses the frequency spectrum to be transmitted to the predominant frequency of half the bit rate. By keeping the modulation index below 0.5 therefore, the bandwidth (in hertz) required is equal to the bit stream rate, and the modulation is referred to as Gaussian minimum shift keying.

The bit rate of 289 350 bit/s cannot be accommodated between adjacent carriers separated by 200 kHz. Allowing a safety margin, a separation of 400 kHz is required, and this means that adjacent carrier frequencies cannot be used in adjacent cells, and imposes an additional restriction on the distribution of carrier frequencies. Fortunately small residual interference signals which will occur will normally be suppressed by the capture effect which is a feature of FM.

The number of bits available out of the total 270 850 bit/s to be divided between the eight channels for traffic purposes depend upon the number of bits required for parity, guard bits, synchronizing, timing sequences, and error correction coding. Each channel allocated to traffic should, after adequate provision for error correction, support a data rate of 12 kbit/s, or any lower rate. Higher rates may require the concatenation of more than one channel.

Speech over GSM has to be digitized and the codec proposed is the RPE-LTP codec, using linear predictive coding. This produces 'toll quality' speech at a bit rate of 13 kbit/s. No error correction need be applied with speech, and it is possible if required to accommodate two speech streams on one channel.

The general principles to be adopted in GSM for locating, paging and routing calls, and also for calling from mobiles and hand-off procedures, are similar to those already in use in TACS. Differences arise due to the digital transmission on the traffic, control and access channels, since these channels no longer have a frequency specific to the channel, but must be identified by the carrier frequency and the slot number of the channel. Also, since the system is designed to operate across national boundaries, an additional 'visitors register' is required to register the mobiles of 'foreign nationality' entering another country.

A more serious problem arising from the use of TDM is that of timing. The bit duration is approximately 3.5 microseconds, and the speed of transmission approximately 3.3 microseconds per kilometre; hence the timing difference between transmitter and receiver (in either direction) represents approximately 1 bit per kilometre. This affects a number of operations:

(a) the relative timing of frequency switching between base and mobile during frequency hopping (see below);
(b) the correct time position of the time slots between base and mobile when measuring channel field strength;
(c) the correct time position of the time slots of a number of mobiles distributed around a cell to ensure that each arrives at the base in its correct position.

This involves the base in monitoring each time slot and advising mobiles on corrective action on their time slot timing.

The TDM system adopted for GSM enables frequency to be changed during transmission provided this is carried out during gaps appearing between bursts. The interchange of carrier frequencies enables the effects of flutter to be minimized. Flutter can be a serious problem and is due to multiple reflections from buildings and obstructions, particularly in built-up areas. Since the pattern varies with frequency a levelling out occurs if the frequencies are interchanged, a process known as frequency hopping. The facility is being incorporated in GSM, and will rely on level measuring at the base station to indicate if and when it should be introduced. The instruction to change frequency at the mobile is made from the base station, and the switching can take place at the frame repetition rate of 217 times per second, ensuring that each channel is switched and that switching takes place at a period between bursts.

4.2.1 Emergency locator service

The FCC in the USA has placed a requirement on mobile phone operators to provide an emergency location service, referred to as E911. The aim is to locate people who need emergency services, but are not able to indicate their location either through injury or through lack of local knowledge. The locator service may also deter hoax calls (or help to catch people making hoax calls) and could have commercial applications in providing directions to businesses such as garages and restaurants.

There are two options: handset-based solutions or network-based solutions. The handset-based solutions generally rely on signals picked up from the Global Positioning System (GPS). Network-based solutions rely on the timing difference in the arrival of signals from neighbouring cell sites. In network-based solutions, the E911 requirement is for the mobile phone to be located within 100 metres at least 67% of the time and to within 300 metres in 95% of cases. In handset-based solutions, the location distances are 50 metres and 100 metres, respectively.

4.2.2 Data over GSM

Standard GSM data service allows a maximum of 9600 baud data rate transmission. An enhanced data rate of 14.4 kbps is also possible with some terminals.

The data link can be either synchronous or asynchronous and the data is conveyed through the network using an ISDN-based data protocol, the ITU-T standard V.110 rate adaptation scheme. Asynchronous transmission requires start and stop bits to be added, so the net data rate is reduced to 7680 bits per second (bps). Synchronous transmission potentially allows a faster data rate, but error checking is employed that requests re-transmission of data when errors are identified and this can produce a lower overall data throughput.

4.2.3 High-speed circuit switched data (HSCSD)

HSCSD allows higher data rates by combining up to four 14.4 kbps data channels together. A maximum data rate of 57.6 kbps is possible.

4.2.4 General packet radio service (GPRS)

GPRS combines up to eight 14.4 kbps channels to give data rates of up to 115.2 kbps. At these rates, Internet browsing is possible. GPRS supports X.25 and IP based transmission, which includes Wireless Application Protocol (WAP) data.

4.3 Other digital mobile systems

Second-generation mobile systems are digital systems. The GSM system has been described in Section 4.2. This section describes the

GSM1800 system, the Personal Communications System (PCS1900) and third-generation systems.

4.3.1 GSM1800 (DCS1800)

The GSM1800 system operates in the 1840–1880 MHz band, with a carrier spacing of 200 kHz. Each carrier gives multiple access through Time Division Multiple Access (TDMA). It is a lower power version of GSM900, which operates in the 900 MHz band. A GSM1800 handset has a maximum power output level of 1 watt, rather than the 2 watts of a GSM900 handset.

Voice coding can be either: half-rate using VSELP at 5.6 kbps, full-rate using RPE-LTP at 13 kbps, or enhanced full rate (EFR) using ACELP at 12.2 kbps. Convolutional coding is used at a rate of $\frac{1}{2}$ to reduce the effects of noise and errors on the radio channel. The radio carrier modulation is Gaussian Minimum Shift Keying (GMSK).

4.3.2 PCS1900 (IS-136)

The Personal Communications System (PCS) is based on GSM and used time division multiplexing to provide separate channels over each radio frequency. Convolutional coding is used at a rate of $\frac{1}{2}$ to reduce the effects of noise and errors on the radio channel. Each carrier is spaced 30 kHz apart and Offset Quadrature Phase-Shift Keying (OQPSK) is used to modulate the carrier.

The voice-coder (vocoder) used by PCS1900 is either Full-Rate using VSELP at 7.95 kbps or Enhanced Full Rate (EFR) that uses ACELP at 7.4 kbps. The EFR scheme was developed for use in PCS1900 and has since been adopted by both GSM900 and GSM1800 networks because it improves the audio quality. Previously, GSM used either a Half-Rate (HR) or Full-Rate (FR) vocoder.

4.3.3 Third-generation systems

The so-called 3 G systems have been described using a number of terms: UMTS, IMT-2000, cdma-2000, w-cdma etc. There are variations between the proposed standards, but in general they all use Code Division Multiple Access (CDMA), convolutional coding, quadrature phase shift keying and time or frequency division duplexing. This technique is described further in Section 4.4.

Third-generation systems will operate in the 1885–2025 MHz band and in the 2110–2200 MHz band. Initially the lower band between 1980 and 2010 MHz will not be allocated, nor will the upper band between 2170 and 2200 MHz. However, the full bands will be available when existing services in these bands can be moved (such as PCS1900, PHS and DECT).

4.4 Code division multiple access – CDMA

Many speech or data channels can share a common path if the transmission system can somehow merge the signals together at the path input in such a way that they can be separated at the path output. The most common techniques are separation in time or frequency. In spread spectrum, the digitized speech or data channel is uniquely coded, so that only a receiver with the same code can restore the speech or data into its original form (by correlation). In fact there are 4.4 trillion possible codes in Qualcomm's system, so the risk of cloning is almost nil. Using a different code for each call allows many calls to be transmitted at the same frequency.

FDMA and TDMA systems switch a call from one cell to another as the caller moves. The connection to the old cell is broken before a new connection is made because the cellphone may have to change its transmit frequency. This is known as hard hand-off. The CDMA system allows connection to be made to the new cell before the old connection is broken because all cells can use the same frequency and no retuning is needed. This type of cell switching is known as soft hand-off. Soft hand-off has the advantage of reducing the mobile's power consumption, reducing interference and increasing capacity.

There is a limit to the number of simultaneous calls; however, CDMA has at least 10 times the capacity of AMPS. The system uses two 1.25 MHz wide frequency bands, one for base station transmissions and the other for mobile transmissions. Adjacent cells use the same frequencies, so frequency planning is simple. It is no longer necessary to allocate individual frequency channels to each cell and have different frequencies in adjacent cells.

The mobile's transmitter power level is adjusted by the base station, so that only the lowest power is used. The base station sends a message to each mobile every 20 ms to find out its power level. If the power level is too high or too low, the base station sends out instructions to increase or decrease the power. Not only does this scheme maximize battery life, it also reduces the risk of interference to other users.

4.5 Digital cordless systems (DECT), CT3

The digital European cordless telephone (DECT) is intended to be the standard European alternative for CT2. As a Pan-European system the standardization process comes under the aegis of ETSI, with a working group known as ETSI-RES3. The advantages over CT2 are intended to be enhanced data handling facilities and interfacing with ISDN, and the ability to receive incoming calls when away from base (CT2 can only receive incoming calls at its fixed terminal or PBX).

The techniques employed are therefore closer to GSM than to CT2. Some simplifications arise from the assumption that the terminal, although moving around the country, is satisfactory during the duration of a call. This removes the need for frequency hopping and hand-off, and the level monitoring required with a moving terminal.

The frequency band allocated for DECT is 1880–1900 MHz. The 20 MHz band is split between 13 radio carriers at 1.7 MHz spacing and each carrier supports 12 TDMA (Time Division Multiple Access) channels. Modulation on speech channels will be GMSK, as already adopted for GSM, but many of the technical details and protocols are still under consideration.

A further variation, known as CT3, has been introduced into the European market, and is claimed to be an advance implementation of DECT. It operates as an extension to a base station or PBX, and also with telepoint stations, but employs TDMA in place of the FDMA used with CT2. It differs from the full DECT recommendations in band allocation and subdivision since it operates in the 800–900 MHz band, using an 8 MHz band divided into eight 1 MHz blocks. Each block carries 16 time slots each of 1 ms duration, providing eight duplex channels. The use of TDMA enables incoming calls to be accepted at telepoints, and also provides a degree of handover between channels, which are facilities that cannot be provided by CT2.

4.6 Personal communications networks

Personal communications networks (PCNs) represent a further step in the quest for an economic personal terminal, ideally pocket sized and capable of dealing with speech or data. No rigid specifications have been established, reliance being placed on competition to produce attractive alternatives to existing systems. It is assumed that the solution will be based on cellular techniques, and proposals so far made appear to favour the GSM approach, although this is not a mandatory requirement.

4.7 Pager systems

Pagers allow someone who is mobile to be sent a message. Cellular handsets are more bulky and cannot always be used; for example, whilst in a meeting or whilst driving. There are three basic services: tone only, numeric display and alpha-numeric display.

The tone only service provides the user with a bleep, flashing light or vibration when the pager number is called. This indicates that someone is trying to contact them, but gives no indication of who that person is. Typically this system is used to tell a busy executive to ring the office.

The numeric display pagers are popular because, using a touch tone phone, it is possible to send a telephone number for the user to call. This is more versatile because any number of callers can leave a message. The pager will bleep when a message is received, the user then looks at the LCD display to find out who is calling. Of course, it is possible to leave a pre-arranged coded message: 1000 may mean phone home, 333 may mean dinner will be ready in half an hour. This type of paging is the most popular.

Alpha-numeric display pagers allow caller to leave a message, e.g. 'phone home' in text rather than as a telephone number or a code. The length of message allowed depends on the pager type but is a minimum of 80 characters, some systems allow up to 2000 characters. The message can be sent by using a computer and modem, or by calling an operator and speaking the message to be sent. Because of these processes, alpha-numeric paging is a small percentage of the paging market.

To avoid a proliferation of codes and systems, the Post Office Code Standardization Advisory Group (POCSAG) was set up, on which the European Selective Paging Manufacturers Association (ESPA) and the Electronic Engineering Association were represented. The objective was a standard code which would allow the transmission of message if required, at a rate of 10 calls per second, and with a capacity of 1 million pagers with up to four alternative addresses per pager. The code finally adopted is known as POCSAG and has been adopted as a ITU-T standard.

POCSAG recommend a bit speed of 512 bit/s with direct FSK applied to the transmitter with a frequency deviation of ± 4.5 kHz (positive deviation representing binary 0 and negative deviation binary 1). The address and the message (when required) are contained in codewords as shown in *Figure 4.3*. The codewords are assembled into frames each comprising two codewords, and a transmitted batch is made up from a synchronizing codeword plus eight frames as shown in *Figure 4.4*.

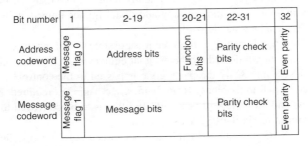

Bit number	1	2-19	20-21	22-31	32
Address codeword	Message flag 0	Address bits	Function bits	Parity check bits	Even parity
Message codeword	Message flag 1	Message bits		Parity check bits	Even parity

Figure 4.3 Codeword format

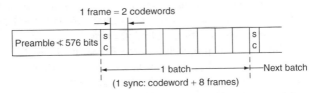

Figure 4.4 Transmission format

Each pager has a 24 bit identity code. The three least significant bits are not transmitted but determine the frame in which the codeword must appear. This means that each pager need only monitor one slot, and can reduce power consumption during the other slots. Bits 20 and 21 select which of the four addresses allocated to the pager are required. The main address is contained in bits 2 to 19. The total capacity is therefore 2^{21} pagers, each of which can have four alternative addresses.

Paging requests are accumulated over a period of about 2 minutes, and assembled into frames and batches, which are then transmitted in a burst. With a message rate of 10 per second the transmitted burst duration is likely to be considerably less than the 2 minute separation between bursts, allowing sequential operation of groups of transmitters if required, and consequent reduction of interference between transmitters.

Each transmission burst is preceded by a preamble consisting of at least 576 bits with 101010... sequence. The synchronizing codeword at the beginning of each batch is 01111100110100100001010111101 1000. The codewords in the remainder of the batch can be either address or message codewords, identified by the first bit (0 for address and 1 for message). Messages frequently require more than one codeword, in which case they occupy additional frames immediately following the address codeword.

If bleep-only paging is required it is possible to dial out the control centre number, followed by the decimal equivalent of the pager code, on a PSTN line. Only the pager code is passed to the control centre, and converted to its binary form. If message paging is required, both address and message will be digital and this is to be conveyed to the control centre via the PSTN using modems.

POCSAG gained worldwide use because it was cheap, efficient and reliable. The only weakness was the error correction code, although alternatives were found to have little if any improvement and were less efficient. To increase the number of pagers in use, higher speeds were needed. Super POCSAG uses the same protocol as POCSAG, but the data rate is increased to 1200 bit/s. POCSAG-2400 increases the data rate still further, to 2400 bit/s.

Other higher speed protocols are the European Radio Message System (ERMES) and the Motorola FLEX protocol. Both of these use synchronous messaging. In synchronous messaging, messages are only sent in a pre-defined time slot. This means that the pager's detection circuits need only be activated part of the time and battery power is saved. This technique also avoids the need for a preamble signal that is sent to synchronize the clock in asynchronous pagers. The time saved in avoiding the preamble allows more messages to be sent from each transmitter, which gives increased efficiency.

ERMES was developed in 1990 and provides a common protocol for all the countries in Europe. The protocol allows for tone, numeric, alpha-numeric and transparent data paging. Multiple radio frequencies are used and each pager scans all the channels. Data is received synchronously by the pager at 6250 bit/s.

FLEX was developed by Motorola in 1993 and allows one of three data rates to be used. The data rate can be 1600, 3200 or 6400 bit/s, and is selected by the company providing the paging service. The FLEX system allows for over one billion different addresses. A two-way pager system, called ReFLEX, allows the pager user to acknowledge a message or send a short reply.

Pagers are usually made as robust clip-on pocket units, using ferrite or loop aerials (which are unfortunately directional). Battery life is an important factor. Message displays, when required, are normally LCD.

Pagers are now being made to fit into portable computers too. Motorola have produced a PCMCIA sized NewsCard; this and an alternative product called AccessCard from Wireless Access enable mobile computer users to receive messages (transparent data messages). These messages could be E-mail text files.

5 Image capture and display

Television is the most popular form of image capture and display system. Generally, images are captured by a camera. The image is focussed onto light sensitive semiconductor material which is then scanned left to right across the image, moving down the image in small steps. This scanning technique is a raster scan. At the display terminal, the screen is scanned in the same raster pattern and is synchronized with the camera. A raster scan is illustrated in *Figure 5.1*.

The camera's image sensor could be a vacuum tube such as a lead oxide vidicon, where the image is focussed onto a lead oxide photocathode. The photocathode emits electrons when photons from the image are received. The electrons are raster scanned and focussed by an electron beam lens, using magnetic fields, onto a cathode. The signal from the cathode is then a voltage proportional to the image luminosity. Colour cameras use three image sensors, one for each of the primary colours: red, blue and green. Colour filters are used in front of each sensor, so that only one colour reaches the photocathode.

Cameras used in close circuit television (CCTV) sometimes use charge coupled device (CCD) sensors. These have an array of semiconductor cells that are arranged in a grid. The charge stored in each cell depends upon the light level focussed onto it. The image is scanned electronically by shifting the stored charge to an output buffer, one line at a time. Thus the output from the CCD is in the same raster format as from the vacuum tube.

As described in Chapter 1, the image is scanned 50 to 60 times per second. Most scanning is interlaced, that is in each frame every other line is scanned. The first scan is of the even numbered lines and the second scan is of the odd numbered lines. In this way the resolution of the picture is maximized whilst the flicker seen at the receiver is minimized. A frame synchronizing pulse is transmitted at the start of each frame, so the raster scan in the camera and display device begin together, and the image is faithfully reproduced.

Display devices are usually cathode ray tubes (CRT). These are similar to the vacuum tube camera, except an electron beam is emitted from the cathode and focussed onto a phosphor screen that glows when struck by electrons. The beam is scanned across the screen in a raster pattern and, as the rate of electrons emitted increases, the display brightens. The number of electrons is proportional to the brightness of the image at the camera. Since the image in the camera and display

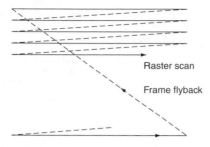

Raster scan

Frame flyback

Figure 5.1 Raster scan

screen are synchronously scanned, the image pattern is reproduced at the display.

In colour CRTs, there are three cathodes ('electron guns'), one associated with each of the primary colours. The screen has phosphor dots that radiate either red, blue or green light when struck by electrons. By inserting a masking screen between the electron guns and the phosphor screen, it is possible to allow only electrons from one gun to reach one colour of phosphor dots. Thus electrons from the blue gun can only reach the blue dots, etc.

Alternatives to the CRT are widely used in portable computers. Thin film transistor (TFT) or liquid crystal displays (LCD) are the most common.

Modern computers display text in the same way as a picture; text characters are converted into a bit-map. Older display equipment had a 'character generator' circuit which only accepted alpha-numeric characters. The character generator was electronically scanned, like a CCD camera device, to output signals at the appropriate moment in the display raster scan.

5.1 Alphanumeric displays – videotext

The obvious extension of the telex printer is to store the received data stream and to translate the coded alphanumeric information into a visual display. This requires a visual display unit (VDU) which provides a scanned raster similar to that used for a television screen. When operated over the PSTN the system is referred to as a videotext. The VDU also contains the character generator required to brighten up the screen in accordance with the stored character information – one character spanning several (typically seven) successive lines to accommodate the height of the character. A block schematic of a typical alphanumeric VDU is shown in *Figure 5.2*.

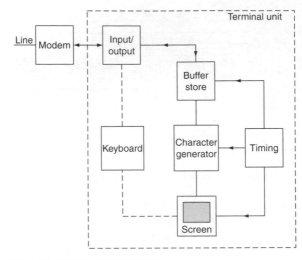

Figure 5.2 Alphanumeric display terminal

An alphanumeric VDU connected to the PSTN or private line can be used for a number of purposes:

(a) To receive incoming information for display.
(b) To store and display messages to be sent to line (for Telex or display) so that they can be checked and edited before transmission.
(c) To obtain and display information on an interactive basis.

Use (b) involves keying into the store and displaying what has been stored. The keyboard is then used to delete, shift, or otherwise edit the displayed text prior to transmission.

Uses (a) and (c) both require outgoing keying to establish the necessary connection (frequently involving a code to gain access and enable appropriate changes to be made). Keying is then required to request the page of information required, and in case (c) to answer questions displayed for an interactive question and answer sequence.

The interactive type of terminal is now widely used by commercial and business users. Banks, building societies, etc., have their own information stores, with access limited to their branches, and usually operating on private circuits. Travel and booking agencies use access to information centres where booking availability can be checked and reservations made. There are also a growing number of independent 'value-added networks' (VANs) providing information of interest to various sectors of the public or business communities, by means of interactive display terminals.

Systems in current use normally employ a 7 bit code. This can be the standard 7 bit International Alphabet No. 5 (IA5) to ISO 646 and ITU-T V3, as adopted for higher-speed telex, or an extended 7 bit code enabling alternative symbol subsets to be selected. The standard symbol set is then referred to as G0, and the alternatives as G1, G2, etc. The alternatives are selected by a control code indicating the start and finish for the subset. The subsets vary considerably with language and application, and are approved and registered by the ISO.

The number of lines of text varies between 12 and 40 (typically 24), with between 30 and 132 character positions per line (typically 80). Approximately 20 000 bits are therefore required for a complete screen display. The telex rate of 300 bit/s is obviously too slow, and operation over a private wire or PSTN line, using a high-speed modem, is necessary to achieve a reasonable display time. A minimum rate of 1200 bit/s is required, and preferably 2400 or 4800 bit/s. This requires FFSK (1200 bit/s) or QPSK (2400 or 4800 bit/s). The higher-speed QPSK is unlikely to be satisfactory over many PSTN lines, but can be used on a suitable private line. The transmission time can be reduced by jumping unused spaces (data compression). This is accomplished by sending a space symbol followed by a numeral indicating the number of unused spaces, but obviously increases the complexity of the processing.

It is apparent that the signals sent from the terminal for interactive operation are normally keyed characters which will not exceed manual keying speed. A duplex modem for such a system can therefore provide high speed in one direction (1200 bit/s or higher) with a lower speed, typically 75 bit/s, in the other direction.

5.2 Video compression

The data rate for video signals is high. Analogue video signals need a bandwidth of at least 6 MHz (black and white) or 8 MHz (colour). Digitizing these signals give data rates of 48 Mbit/s or 64 Mbit/s, respectively, assuming 8-bit analogue to digital conversion.

When video signals are analysed it is apparent that, from one frame to the next, the amount of change in the image is small. Quite often the background remains unchanged while a presenter talks to the camera, thus only slight head movements take place. The amount of data that needs to be transmitted for these changes is a fraction of the 64 Mbit/s needed for the whole image.

Video-conferencing systems are designed on the principle that the background does not change. In a conference, the background is an office. Only the difference between one image and the next is transmitted. Overall, the data rate is reduced to about 1.28 Mbit/s. Low data rate systems are produced for videophone applications. Here, the refresh rate and the picture resolution are lower than those used in broadcast television and data rates as low as 64 kbit/s are possible.

It is possible to compress video films for later transmission. At present prerecorded films and programmes are processed frame by frame. There are two compression techniques known after their ISO group name: Joint Photographic Experts Group (JPEG) and Moving Pictures Expert Group (MPEG). JPEG is designed for compression of still images and will not be described further. MPEG is designed for compression of motion pictures. Real time (live) compression of digitized pictures is now possible using high-speed digital signal processors (DSPs).

Using MPEG encoding, the data rate is reduced to 1.5 Mbit/s. Pictures are compressed in two stages. First, by converting the image into a number of 8×8 pixel blocks and then taking a discrete cosine transform (DCT) of each block. Then a motion predictor is used to estimate what the next image will be. The result is simplified by mathematical processing to produce a data stream of digital 1s and 0s. The data stream is then compressed by Huffman coding using fixed tables, which looks for long sequences of 1s and 0s and reduces them to a simple code (see Chapter 7 on facsimile data compression).

Slow-scan television does not compress the video image. Instead the image is stored in the camera, transmitted slowly and then the displayed image is updated. The display terminal also has a store so that the picture is updated in a single frame, rather than gradually as the data is received.

6 Modems

6.1 Overview

A MODEM (MOdulator–DEModulator) is fitted to each end of a data link to change the serial data signals produced by the data source into voice-frequency signals suitable for transmission over the telephone network (PSTN). A modem also converts voice-frequency signals to serial binary data for onward transmission to a data terminal. A modem may operate in one or more of three modes:

1 Simplex – in one direction of transmission only.
2 Half duplex – can operate in either direction but in only one direction at a time.
3 Full duplex – can operate in both directions simultaneously.

Because parts of the frequency spectrum of the PSTN speech circuits are used for signalling and supervisory functions only the bands 300–500 Hz and 900–2100 Hz are available for data transmission. Low-speed data systems (up to 1200 bit/s) normally use frequency shift keying in which 1 and 0 are represented by different audio frequencies. In higher-speed modems (2400 bit/s and over) the data is encoded before modulation in order to reduce the effective bit rate of the transmission. One technique is to pair bits into the four possible combinations 00, 01, 10 and 11 which are then called DIBITS; at the cost of increased complexity the process can be extended so that a modulation change occurs only once for every 4 bits of information transmitted.

Several modulation techniques are used (see also Chapter 12) in addition to frequency shift already mentioned; these include phase modulation, differential phase modulation, quadrature amplitude modulation and vestigial sideband amplitude modulation.

Modems must be approved by the local government authority and must provide isolation between the PSTN and other apparatus that may carry unsafe potentials, for example, 240 V 50 Hz.

6.2 Frequency shift key (FSK)

V.21 (or Bell 103) modems transmit data at speeds up to 300 bit/s in both directions simultaneously (i.e. full duplex). These modems are quite simple in operation since they use fixed frequency tones to signal a logic 1 or a logic 0. In fact, the two ends of a system are configured so that the tones transmitted from one end are different from the tones transmitted from the other end. This allows data to be transmitted from both ends simultaneously since the frequency spectrum used by one end is separate from the other end. If there were any overlap in the frequency spectrum used by any of the signals, simultaneous transmission would have been impossible.

The ITU-T V.21 standard defines the transmit frequencies (*Table 6.1*):

Table 6.1

Channel	Mean frequency (Hz)	Logic 1 (Hz)	Logic 0 (Hz)
1	1080	980	1180
2	1750	1650	1850

Data transmission over radio links is reliable with V.21 modems. One of the few uses left for V.21 modems used on copper pairs is the Telex network.

The requirement for V.23 (Bell 202) modems is almost non-existent now, however modems that could operate faster than V.21 were required for the operation of dumb terminals working off main-frame computers. The V.23 standard provides for a maximum data rate of 1200 bit/s. However, to allow for poor analogue links between the modems, a fall-back rate of 600 bit/s is provided. The keyboard provided the data to be transmitted back to the computer and, since this tends to be quite slow, a maximum data rate of 75 bit/s is provided.

The ITU-T V.23 standard assigns frequencies as shown in *Table 6.2*.

Table 6.2

Channel	Mean frequency (Hz)	Logic 1 (Hz)	Logic 0 (Hz)
1 (600 bit/s)	1500	1300	1700
1 (1200 bit/s)	1700	1300	2100
2 (75 bit/s)	420	390	450

6.3 Phase modulation

Modems using phase modulation usually employ the technique of 'differential phase modulation' whereby changes in phase rather than absolute phase determine a binary 1 or 0. So a given phase change would be a '1' if the preceding digit was '0' and vice versa. This avoids the necessity to generate a reference signal of constant phase which would be required if '0' and '1' were allocated fixed phase positions.

Common practice is to use four phase changes, each one of which represents 2 bits of data (i.e. a dibit), and this reduces the bandwidth required for transmission since the maximum fundamental frequency of the dibit waveform is $\frac{1}{4} \times$ bit rate. The protocols using four phases to indicate dibits are ITU-T V.22 (Bell 212) and V.26 (Bell 201).

The V.22 modem allows a full duplex data transmission rate of 1200 bit/s and, since one symbol is transmitted for every two data bits, the symbol rate is 600 baud. The originating modem uses a 1200 Hz carrier to transmit data. The answering modem uses a 2400 Hz carrier to transmit data in the reverse direction.

V.26 modems have a maximum data rate of 2400 bit/s, with a fall back rate of 1200 bit/s. To raise the data transmission rate to 2400 bit/s, phase modulation of an 1800 Hz carrier is used. What is more, each symbol transmitted to line represents two data bits. The maximum baud rate is therefore 1200 baud, with a fall-back to 600 baud.

There are two versions of V.26, alternative A and alternative B. Alternative A uses phase changes of 0, 90, 180, or 270 degrees for each pair of bits. Alternative B uses phase changes of 45, 135, 225, or 315 degrees for each pair of bits. Operation over the public switched network requires the modem standard V.26 bis to be applied. V.26 bis uses alternative B phase modulation. Operation over a 4-wire private circuit can use either alternative A or alternative B phase modulation (alternative A is most commonly used for V.26).

Table 6.3

Dibits	V.22	V.26 (alternative A)	V.26 bis (alternative B)
00	90	0	45
01	0	90	135
11	270	180	225
10	180	270	315

An 1800 Hz carrier is typical for this form of modulation since the group delay versus frequency distortion of a telephone line is small at

46

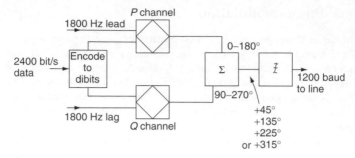

Figure 6.1

this frequency. A simplified block diagram of a modem using V26A
four-phase modulation is shown in *Figure 6.1*. Here serial 2400 bit/s
data is held in a 2 bit shift register so that dibits are generated at
1200 bit/s. The first bit of a pair is channelled to the P modulator and
the second bit to the Q modulator. The P modulator is fed with an
1800 Hz carrier which leads the Q modulator carrier by 90°. The two
phase-modulated outputs are summed, and fed to line via a low-pass
filter.

At the receiving modem the 1800 Hz carrier is recovered (at 0°
and 90°) using a locally generated signal phase-locked to the received
signal and the phase changes detected by the P and Q polarity detectors
which in turn generate translation in the decoding logic back to the
original 2400 bit/s data. Clearly, accurate synchronism between send
and receive modems is essential for the correct detection of phase
changes of incoming signals.

Since angular velocity ($\omega = 2\pi f$) is the time derivative of phase,
then phase modulation by a logic signal produces a change in instantaneous carrier frequency. The actual frequencies generated for each
dibit for V26A and V26B systems are (for a 2400 bit/s data) shown
in *Table 6.4*.

Table 6.4

Dibit	00	01	10	11
Alternative A	1800	900	1500	1200
		2100	2700	2400
Alternative B	750	1050	1650	1350
	1950	2250	2850	2550

The bandwidths required are $(2700 - 900 = 1800 \text{ Hz})$ and
$(2850 - 750 = 2100 \text{ Hz})$ which is less than would be needed for a

2400 bit/s FSK system (i.e. 2400 Hz). The relative amplitudes of these spectral frequencies have an envelope which follows a $(\sin x/x)^2$ envelope as shown in *Figure 6.2* for repetitive dibits. When the modulation pattern is complex the energy is spread more uniformly across the spectrum.

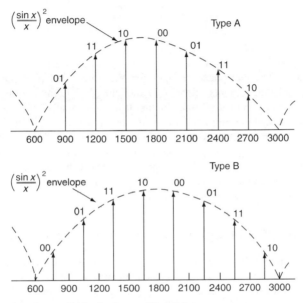

Figure 6.2 Energy distribution for repetitive dibits

For minimum intersymbol interference in any isochronous digital transmission, filtering must be introduced to produce a roll-off of the energy spectrum about the Nyquist frequency (i.e. half the modulation rate in bauds). In the above example this is 600 Hz above and below the carrier frequency. Filtering at the transmitter may only remove the higher-order lobes of the spectrum so as not to lose timing information in the transmitted data stream; this is particularly the case with dibits 00 and 10 in type B modulation.

Increasing the data rate to 4800 bit/s is achieved in the ITU-T V.27 protocol by letting three bits equal one baud. The baud rate is thus 1600 baud. Eight possible states can be realized using three data bits, so V.27 requires eight phases of an 1800 Hz carrier. The reduced phase margin requires the line to be equalized, that is adjusted to reduce phase shifts in the line.

Table 6.5

Data bits	Phase change
001	0
000	45
010	90
011	135
111	180
110	225
100	270
101	315

The differential phase changes for V.27 are listed in *Table 6.5*. There are three variants of the V.27 modem. V.27 is for use on high quality voice band leased line circuits. It has a manually set equalizer that is adjusted during installation and commissioning. V.27 bis is for use on general quality voice band leased line circuits. The equalizer automatically adapts itself to the line conditions. Finally, V.27 ter is a standard for use on public switched networks. It has automatic line equalization and a fall-back to V.26A (i.e. a rate of 2400 bit/s) if line conditions do not allow the higher rate.

6.4 Multilevel signals and the Gray code

The telegraph printer referred to above is a simple example of the use of a two-level signal ($\pm v$). This principle can be extended to a general case of m levels in which case the digital signal is described as having a signal alphabet of value 'm'. The term 'level' does not necessarily refer to amplitude; it may be frequency, phase or amplitude or, indeed, a combination of two of these.

If we put $m = 2^n$ (i.e. $n = \log_2 m$) the m-level data symbol is represented by 'n' binary digits each of which may be '0' or '1'. So, for example, an eight-level signal has $n = 3$ giving the binary sequences 000, 001, 011, 010, 110, 111, 101 and 100 for the eight levels. Note that these differ by only one digit for each transition between adjacent levels; this coding system for binary signals is known as the Gray code. In these conditions the errors in detection of an m-level element, which arise at the crossing of a decision threshold, are likely to give the minimum error to the recovered binary waveform.

In a multilevel signal where each element contains 'n' bits ($n = \log_2 m$), and the signal element rate is B bauds (elements/second), the transmission rate will be $\log_2 m$ bit/s.

6.5 Quadrature amplitude and trellis coded modulation

To reliably increase the data rate above 2400 bit/s, it is necessary to increase the complexity of the signal transmitted to line. Rather than just use phase shifts of a carrier, which would require the detection of very small phase differences, both phase and amplitude are used.

Quadrature amplitude modulation (QAM) uses differential signalling so that the receive modem compares the amplitude and phase of a signal during the transmission of one symbol relative to the previous symbol. The constellation diagram in *Figure 6.3* now shows how both the amplitude and the phase contribute to the signal. The term 'signal space' is sometimes used to describe the constellation. Each symbol must have its own signal space since amplitude and phase noise will cause the signal to be in the general area of a constellation point.

Figure 6.3 shows the vector positions obtained, the tips of which form the dots of the so-called constellation CRT display used for the evaluation of vector modulation.

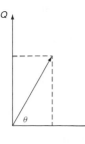

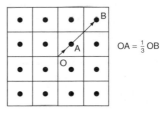

OA = $\frac{1}{3}$ OB

A QAM signal element vector diagram

Four amplitude level, four phase QAM vectors

Figure 6.3

In *Figure 6.3* any QAM signal, $F(t)$, is given by

$$F(t) = I(t)\cos(\omega_c t + \phi) + Q(t)\cos\left(\omega_c t + \frac{\pi}{2} + \phi\right)$$

where $I(t)$ and $Q(t)$ are the in-phase and quadrature modulation waveforms. For an interval t over which $I(t)$ and $Q(t)$ may be deemed constant,

$$F(t) = I\cos(\omega_c t + \phi) + Q\cos\left(\omega_c t + \frac{\pi}{2} + \phi\right)$$

$$= k\cos(\omega_c t + \theta + \phi)$$

where

$$k = \sqrt{I^2 + Q^2} \text{ and } \theta = \tan^{-1}(Q/I)$$

In a 64 level QAM system, use is made of adaptive equalization in the receiver to compensate for line attenuation which may vary with time. The demodulated baseband signal at the receiver is used to adjust the tap gains on a digital filter to make it appear as the inverse of the channel response from the transmitter.

Trellis coded modulation uses a combination of QAM and 'convolutional coding.' In simple terms the phase and amplitude of each symbol is determined by looking at the data stream; not just the bits used for the symbol in question, but those used for previous and future symbols too. Thus the sequence of phase and amplitude changes is indirectly determined by the sequence of data. Some sequences of constellation point moves are not allowed, in order to improve the error checking.

The V.22 bis standard allows transmission at 2400 bit/s; the baud rate is 600 baud. Two carrier frequencies are used, as with V.22, so that transmit and receive paths can be separated by filtering. The carrier frequencies are 1200 and 2400 Hz. However, instead of a four point constellation there is a sixteen point constellation, four points in each quadrant. Each point in the constellation represents four data bits. Two bits of data control the amplitude and the other two bits control the phase.

The V.32 modem increases data throughput to 9600 bit/s. It uses either QAM signalling with a 16 point constellation, or trellis coded modulation (TCM). The baud rate is 2400 baud and the carrier frequency is 1800 Hz.

The V.32 bis modem uses more constellation points to raise the data rate to 14.4 kbit/s. The baud rate is 2400 baud and the carrier frequency is 1800 Hz. A maximum of six data bits per symbol must be encoded, which requires a constellation with 128 points. Fall-back rates are 12 kbit/s (5 bits per symbol), 9.6 kbit/s (4 bits per symbol), 7.2 kbit/s (3 bits per symbol) and 4.8 kbit/s (2 bits per symbol); TCM is used for all data rates.

The V.32 ter allows a maximum data rate of 19.2 kbit/s. The baud rate is 2400 baud and the carrier frequency is 1800 Hz. Fallback rates of 16.8 and 14 kbit/s are provided. TCM is used for all data rates.

The V.34 modem is the ultimate analogue modem, with a limiting data rate of 33.6 kbit/s. This rate is believed to be the maximum possible data rate over a public switched telephone network (PSTN). The limitation is due to two factors: (1) the bandwidth allowed through the analogue to digital converter in the telephone exchange is 300 Hz to 3.4 kHz; (2) the converter uses either an 8-bit A-law or a mu-law encoding scheme that gives 38 dB signal to noise ratio for signals in the −40 dBm to −15 dBm amplitude range. The bandwidth limitation means that the symbol rate (baud rate) is limited. The signal to noise ratio limits the number of signal levels that can be differentiated.

Strictly, V.34 is limited to 28.8 kbit/s, and it is V.34 plus which reaches the dizzy rate of 33.6 kbit/s. Unlike other modems, both the baud rate and the carrier frequency depend upon the line conditions. The two connected modems transmit test tones to each other in order to analyse the line conditions. The baud rate can rise to 3429 baud under good conditions. The modulation scheme is TCM.

Two 'modem' standards not described here are V.90 and V.92. These use completely different transmission techniques and are described in Chapter 8.

6.6 Modem delay time

Delays of the order of a few milliseconds are introduced by filters, equalizers, etc., within a modem, and propagation times through the telephone network can be 10–20 ms within the UK but over 0.5 s via satellite on an intercontinental link. High-speed modems (>2400 bit/s) may incorporate adaptive equalizers for echo cancellation and these may require set-up times of 100 ms to several seconds.

A source of delay, even on short routes with low-speed modems is the 'turnaround' time of a modem, for example, the interval between 'Request to Send' and the reply 'Ready for Sending'.

6.7 Echo suppression and cancellation

Formulae for characteristic impedance, loss, phase shift, velocity, etc., of four-terminal networks and infinite lines are given in Chapter 26.

Terminal apparatus is rarely a perfect impedance match to the line, and likewise mismatches occur wherever hybrid transformers are used, for example to convert two-wire to four-wire working. The net effect is that reflections occurring at mismatches travel back to the transmitting end so the speaker hears an echo of his or her own voice. On short lines this is not a problem since it appears virtually as sidetone (see Chapter 2) but on international links suppressors are fitted to the line repeaters in the four-wire circuit. These operate by detecting the repeater output of the transmit line and using it to reduce the gain of the 'receive' line repeater at the same point. During data transmission,

52

between modems for example, it may be necessary to have full duplex working so that the return channel can send control signals. Means have to be provided to disable the echo suppressors in this case.

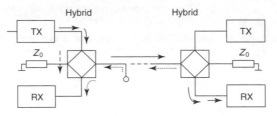

Figure 6.4 Transmission path, →; near-end echo, – →; far-end echo, · · · →

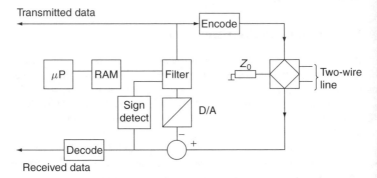

Figure 6.5

Echo cancellation techniques can be used for duplex data transmission between modems on a two-wire baseband circuit. The basic principle involves digital signal processing techniques whereby an adaptive transversal filter simulates the echo path from transmitter to receiver, thus allowing the echo signals to be subtracted from the received signal. Changes in line length and impedance mismatches can be accommodated by changing the filter coefficients stored in the RAM under microprocessor control. *Figure 6.4* shows the route by which near- and far-end echoes reach the receiver and *Figure 6.5* the schematic of an adaptive filter echo cancellor. Typical attenuation through a hybrid for a near-end echo is 10 dB while line attenuation may be 40 dB more, so the echo level signal is 30 dB above the incoming signal. Far-end echo is a much lower-level signal but delayed in the order of milliseconds.

The microprocessor which determines the digital filter coefficients is controlled by an algorithm designed to minimize the error in the received signal. Various algorithms have been studied and the results obtained on a trial in Norway showed that performance met the requirements of a BER of 10^{-7} over 15 km of 0.6 mm cable where attenuation was 50 dB at a data baud rate of 12 kbit/s. Echo cancellation convergence (i.e. filter adjustment time) was $2-3$ s after power on.

6.8 Security

The increasing use of national and international public communication networks, for the exchange of sensitive and confidential data, for example financial information, forensic evidence, technical design data, has created the need for security procedures to make it impossible for any unauthorized person to gain access to such information. The evolving OSI (Open Systems Interconnection) will give even more prominence to this problem and ISO and ITU-T recommendations on the subject may be expected.

Unauthorized access has been reported via telephone lines into electronic mail, facsimile and computer links, and also via a radiofrequency scanner positioned near a building containing computer and facsimile terminals. Such access (or 'hacking' as it is generally known) to facsimile machines has in fact been facilitated in advanced design fax terminals by virtue of their remote programming capabilities, which the hacker can reprogram to advantage, thereby, for example, receiving copies of all transmitted files from the targeted machine.

A simple but relatively low level of security against access to a machine's memories and instruction sets is offered by a 'password' – a mnemonic or code number known only to authorized personnel which must be used to 'unlock' the machine and make it ready for use. An everyday example of this, albeit in a different context, is a four-digit code which must be used to reactivate a car radio once it has been disconnected from the battery. Offering 9999 possible combinations it also includes a timer so that each wrong attempt is separated by a time interval starting at 1 minute and increasing in a geometric ratio of 2.

The principal method used to provide security is that of encryption whereby data is converted, before transmission, into a cipher text that is unreadable by an unauthorized person. The transformation from data text to cipher text is performed by an algorithm under the control of a key, the security of which is paramount. This is the unique difference between encryption and the scrambling of data bits to reduce the probability of correlation between received and transmitted data subsequently encoded for line transmission.

There are two main categories of encryption: symmetrical and asymmetrical. In the former case the identical key can be used for encryption at the transmitter and decryption at the receiver, whereas in asymmetrical encryption the two processes require different keys.

The algorithm for symmetrical encryption should ensure that:

(a) it cannot be deduced from a study of the data and cipher texts;
(b) with a knowledge of the key, encryption and decryption processes are simply performed;
(c) it is neither practically or economically feasible to deduce the relation between data text, cipher text and key because the total number of possible combinations is so large.

Very large prime numbers have been used in this context since they have so far been impossible to calculate except on a supercomputer.

In asymmetrical encryption, the actual encryption key can be released for general use within an organization, but only a recipient with the decryption key can recover the data text.

The distribution of cipher keys within a security-conscious organization is obviously a prime consideration; one possibility is to use a key distribution centre employing only strongly security-vetted personnel following a strict procedure. For example, if terminal A wishes to send an encrypted message to terminal B, A first identifies itself to the key distribution centre requesting a key to communicate with B. The centre communicates a reply to A including a key enciphered under both A and B's unique masterkeys whereupon A deciphers this message, retaining the communication key, and forwards the remainder, plus additional encryption initiation data generated by A, to terminal B. Terminal B authorizes this encryption initiation data with the actual time of receipt, encrypts it and returns it to A. These message exchanges establish a common communications key at A and B allowing encrypted data exchange to proceed.

6.9 Error control and correction

Clearly errors in the form of missing or wrong digits can cause significant problems in digital data transmission, and even in speech transmission may result in a lack of synchronization with subsequent distortion of the decoded analogue speech signal. Various systems of error detection and control are in common use, as described below.

6.9.1 Vertical redundancy check (VRC)

A parity bit is added to each character. ITU-T Recommendation V4 is that the least significant bit is transmitted first and the parity bit last in each character block.

EVEN PARITY occurs when the total number of 1s in the block, including the parity bit, is even.

ODD PARITY occurs when the total number of 1s in the block, including the parity bit, is odd.

VRC obviously detects single-bit errors in a character block, but fails if 2 bits are incorrect. Its use is generally restricted to the peripherals of a single computer where single-bit errors are most likely to occur.

6.9.2 Longitudinal redundancy check (LRC)

An extra byte is transmitted at the end of a block of characters to form a parity character, also called the block check character (BCC). The first bit of the BCC 'samples' the first bit of each data character in the block, acting as a parity bit; the second bit of the BCC acts as a parity bit for all the second bits of each data character in the block; and so on. BS 4505 Part III specifies even parity for the LRC.

By combining VRCs and LRCs, all odd numbers of errors plus 2 bit errors can be detected, but since many errors are due to induced impulse-type noise, some errors still remain undetected. The trade-off against error detection is in increased transmission time; for example, for VRC 1 bit in 8 of an ASCII set is usually for parity.

6.9.3 Cyclic redundancy check (CRC)

This system of error detection involves dividing the digital stream of characters by a specified binary number. ITU-T Recommendation V41 uses the polynomial $x^{16} + x^{12} + x^5 + x^0$ (in binary notation 10001000000100001), and a data block size of 260 bits including service and check bits. The remainder from this division is transmitted as a series of check digits after the data block. At the receiving end the data block plus remainder is divided by the same polynomial; there will be no remainder from this division in an error-free

56

transmission. Such a scheme can improve a typical error rate of 1 in 10^4 to 2 to 10^9.

6.9.4 Error correction systems

The detection systems described above are made to send an automatic request for repetition (ARQ) back to the transmit station. Two methods of correction are in common use:

1 *Stop and wait:* After each data block, the transmitter stops and waits for a positive acknowledgement (ACK) before sending the next block. If it receives ARQ instead, the previous block is repeated. This is not efficient on satellite links because of the relatively long propagation time involved.
2 *Continuous retransmission:* The transmitter has sufficient buffer storage to store a succession of transmitted data blocks. Each ACK received deletes the relevant data blocks from the store whilst each ARQ causes the appropriate block to be retransmitted.

6.9.5 Forward error control (FEC)

Several codes have been devised that will not only detect but also automatically allow for correction of errors since they identify the position of the error bit(s) in the data field. Such codes are attributed to Hamming, Bose, Chaudhuri and others. The Hamming code is now described in principle as an example.

Check bits are inserted in positions corresponding to 2^n in a data field. For example,

15	14	13	12	11	10	9	8	7	6	5	4	3	2	1
D	D	D	D	D	D	D	X	D	D	D	X	D	X	X

where D = data and X = check bit in positions 2^0, 2^1, 2^2 and 2^3 (i.e. 1, 2, 4 and 8 respectively). So a typical message could be:

1 0 1 0 1 0 1 X 1 0 0 X 1 X X

and the bit positions containing a data '1' are added in modulo 2:

15	1	1	1	1
13	1	1	0	1
11	1	0	1	1
9	1	0	0	1
7	0	1	1	1
3	0	0	1	1
	0	1	0	0

The modulo 2 sum bits are used as check bits so the transmitted signal becomes

$$1 \quad 0 \quad 1 \quad 0 \quad 1 \quad 0 \quad 1 \quad 0 \quad 1 \quad 0 \quad 0 \quad 1 \quad 1 \quad 0 \quad 0$$

At the receiving end, the binary values of each bit position containing a '1' are added:

15	1	1	1	1
13	1	1	0	1
11	1	0	1	1
9	1	0	0	1
7	0	1	1	1
4	0	1	0	0
3	0	0	1	1
	0	0	0	0

(Zero sum indicates error free.) If there is a single-bit error, for example if position 15 receives a '0' instead of a '1':

13	1	1	0	1
11	1	0	1	1
9	1	0	0	1
7	0	1	1	1
4	0	1	0	0
3	0	0	1	1
15	1	1	1	1

(The denary value of the sum is 15.) The sum now shows bit 15 is an error and this is automatically inverted to a '1' in the receiver.

This particular code will correct single-bit errors and detect double errors but may fail to detect some multiple errors.

The Bose–Chaudhuri codes form blocks of 2^{n-1} bits in such a way that q errors can be detected by the inclusion of nq parity bits. Hence for $n = 5$, a block of 31 bits with 21 message bits and 10 parity bits can correct double errors and will detect quadruple errors.

6.10 Transparency

In the context of data transmission, the ability of a system to handle all possible sequences of data bit values is referred to as transparency, and a transparent text is one in which any combination of code can be used.

In a system where the receive terminal extracts the timing wave-form essential for data demodulation from the data stream itself, it is necessary to restrict the length of sequences of data bits during which there are no timing transitions. This will therefore not be a transparent system.

Transparency in data exchange has to ensure that characters such as ACK, NAK, EOT, etc., which are used for line set-up and control, can be used in the data text as well, but when they are in text they are ignored by the data receiver (modem). One method is to insert data link escape (DLE) characters in a message to indicate that a number of following adjoining characters are to be ignored by the line control circuit in the receiver.

An alternative method (synchronous data link control (SDLC)) dispenses with conventional control characters and instead uses the message format

$$F - A - C - \text{Text} - BC - F$$

where F is a control frame always 01111110, A is an 8 bit address, C is an 8 bit control and BC is the ITU-T V41 cyclic code polynomial $x^{16} + x^{12} + x^5 + x^0$.

6.11 Modem connections

Connections between the computer or VDU terminal (i.e. the data terminating equipment – DTE) and the modem or data circuit equipment (DCE) are known as the 100 series interchange circuits and are defined in ITU-T Recommendations V24 (*Table 6.6*).

The connectors used with 100 series interchange circuits and their pin-out are defined by ISO 2110 and for modems in accordance with ITU-T V21, V23, V26, V26*bis*, V27 and V27*bis* are as shown in *Table 6.7*.

6.12 RS232 interface

The RS232 interface is almost universally used to connect computers to modems. The signals are unbalanced, that is they are referenced to a 0 V or ground terminal. The data and control signal levels are bipolar. For the control signals, a logic 1 (ON) has a positive voltage of between +5 V and +15 V, and a logic 0 (OFF) has a negative voltage of between −5 V and −15 V. Modern versions of the RS232 standard, RS232D and RS232E, the ON and OFF voltage ranges extend to 25 V.

Unusually the data circuits in RS232 have the reverse polarity and are limited to 15 V. A logic 1 can have a negative voltage between

Table 6.6

No.	Interchange circuit — Description	Data From DCE	Data To DCE	Control From DCE	Control To DCE	Timing From DCE	Timing To DCE
101	Protective ground/earth						
102	Signal ground or common						
103	Transmitted data		•				
104	Received data	•					
105	Request to send				•		
106	Ready for sending			•			
107	Data set ready			•			
108/1	Connect data set to line				•		
108/2	Data terminal ready				•		
109	Data channel receive line signal detector			•			
110	Signal quality detector			•			
111	Data signalling rate select (DTE)				•		
112	Data signalling rate select (DCE)			•			
113	Transmitter signal timing (DTE)						•
114	Transmitter signal timing (DCE)					•	
115	Receiver signal timing (DCE)					•	
116	Select standby				•		
117	Standby indicator			•			
118	Transmitted backward channel data						
119	Received backward channel data	•					

(continued overleaf)

Table 6.6 *(continued)*

No.	Interchange circuit Description	Data From DCE	Data To DCE	Control From DCE	Control To DCE	Timing From DCE	Timing To DCE
120	Transmit backward channel line signal						
121	Backward channel ready			•	•		
122	Backward channel received line signal detector			•			
123	Backward channel received line quality detector			•			
124	Select frequency groups			•	•		
125	Calling indicator						
126	Select transmit frequency				•		
127	Select receive frequency				•		
128	Received signal element timing (DTE)						•
129	Request to receive				•		
130	Transmit backward tone				•		
131	Received character timing					•	
132	Return to non-data mode				•		
133	Ready for receiving				•		
134	Received data present			•			
191	Transmitted voice answer				•		
192	Received voice answer			•			

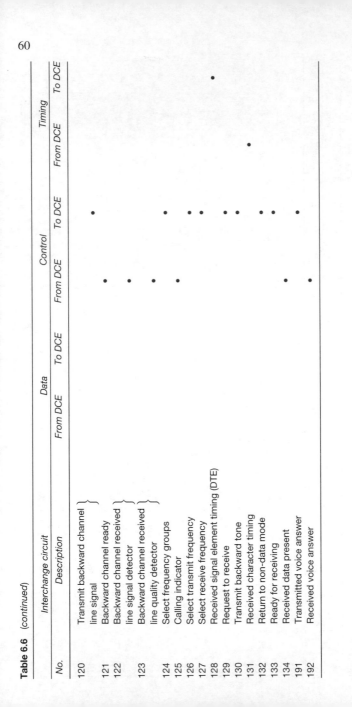

Table 6.7

Pin no.	V21	V23	V26/27
1	*	*	*
2	103	103	103
3	104	104	104
4	105	105	105
5	106	106	106
6	107	107	107
7	102	102	102
8	109	109	109
9	N	N	N
10	N	N	N
11	126	N	N
12	F	122	122
13	F	121	121
14	F	118	118
15	F	2	114
16	F	119	119
17	F	2	115
18	141	141	141
19	F	120	120
20	108/1-/2	108/1-/2	108/1-/2
21	140	140	140
22	125	125	125
23	N	111	111
24	N	N	113
25	142	142	142

* Pin 1 is assigned for connecting shields of shielded cables; it may be connected to protective ground or to signal ground
F: reserved for future use
N: reserved for national use

−5 V and −15 V, whilst a logic 0 can have a positive voltage of between +5 V and +15 V. The RS232 interface is designed to work over 15 metres of cable, which is adequate for most applications.

There can be confusion about the individual wire connections because of the terminology used and because it is a full duplex interface. Also, there are two connectors used for RS232: 9-pin and 25-pin. The 9-pin connector is popular because there is limited space on small equipment enclosures and special adaptors are available to convert from 9-pin to 25-pin and vice versa.

The computer is known as the data terminal equipment (DTE). The pin labelled 'transmit data' or T × D, is used as an output from the computer. This is intuitively logical. The modem is known as the data communications equipment (DCE). The pin labelled 'transmit data'

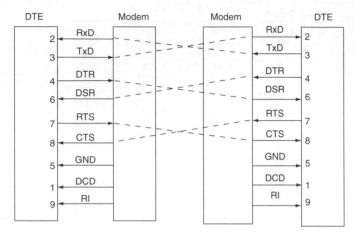

Figure 6.6 RS232 network connections (9-pin connectors)

Figure 6.7 Null modem (9-pin connectors)

is used to receive data from the computer for transmission over the telephone network. This is where the confusion sometimes arises – a transmit signal is an input.

Other pins in the RS232 connector can be similarly confusing. *Figure 6.6* may help in the understanding.

If two computers need to be interconnected a 'null modem' is required. This is an adaptor that crosses over the transmit and receive connections. Without this adaptor the transmit connector on one computer would be connected to the transmit connector on the other computer. Not only would data transfer fail, some damage could be caused to the RS232 drive circuits. A diagram of the null modem is given in *Figure 6.7*.

The pin assignment for 9-pin D-type connectors is given in *Table 6.8*.

Table 6.8 RS232 connector pin assignments (9-pin connector)

Pin no.	Pin name	Description	Modem in/out
1	DCD	Data carrier detect	Output
2	R × D	Receive data	Output
3	T × D	Transmit data	Input
4	DTR	Data terminal ready	Input
5	GND	Ground	
6	DSR	Data set ready	Output
7	RTS	Ready to send	Input
8	CTS	Clear to send	Output
9	RI	Ring indicator	Incoming call set up

7 Facsimile

Telex provides a satisfactory means of transmitting typescript, but does not provide an actual copy of a document. It cannot therefore print signatures or diagrams. Facsimile (FAX) provides an accurate copy, and therefore has a wider range of application than telex. A typical terminal is shown in block schematic form in *Figure 7.1*.

The document to be transmitted is scanned horizontally, line by line; hence the closeness of the lines determines the vertical resolution (for reasonable-quality resolution this is normally 3.5 to 4 lines per millimetre). The vertical displacement to space the lines is relatively slow and conveniently provided by slow mechanical advance of either the paper or scanning head. The horizontal scan must produce an analogue or digital signal representing the reflected light intensity along the line. This can be provided mechanically by traversing the scanning head horizontally across the page for each line, but this requires a very rapid return of the head between each line scan. Alternatively mechanical scanning can be obtained by wrapping the paper around a drum. One line is then scanned for each revolution of the drum, the scanning head advancing one line spacing for each revolution. This provides a simple robust mechanical arrangement, but does not enable paper rolls to be used. The mechanically scanned systems employ a photoelectric transducer, focused on to the paper, to obtain the electrical signal, and various optical systems have been devised to provide the horizontal scan, such as the rotating polygon of mirrors. Each mirror in turn provides an optical path between a point on the paper and the photoelectric transducer, the point traversing across the paper as the mirror rotates.

No advantage can be gained by using electronic line spacing; in fact, mechanical movement of the paper is required for paper feed with both sheets and rolls. Considerable advantage can, however, be obtained by using electronic horizontal scanning, which removes the speed restrictions imposed by a mechanical system. This enables very rapid flyback to be achieved and also makes possible a rapid scan to reduce transmission and printing time.

Electronic scanning can be carried out either by deflecting an electron beam to scan an optical image of the line, or by an array of photoelectric elements which monitors the whole line simultaneously and is extracted as a serial signal.

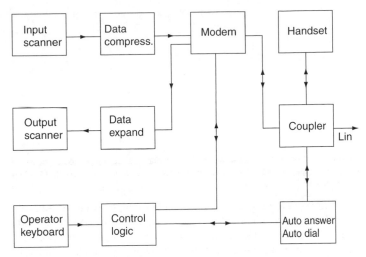

Figure 7.1 Facsimile terminal

The horizontal resolution is controlled by the number of picture elements (pixels) per millimetre. A picture element is the minimum distance in which the change white to black and black to white can be achieved, which in a digital system corresponds to the distance represented by 1 bit. Hence for equal vertical and horizontal resolution the pixel width or the bit width should equal the line spacing, for example a line spacing of four lines per millimetre would require 16 pixels per square millimetre for equal vertical and horizontal resolution, and transmission would require 1600 bits per square centimetre of document.

A distinction must be drawn between facsimile in which only black or white reproduction is required, and facsimile capable of reproducing graduated 'grey' tones. For most document reproduction including signatures and line diagrams, black and white printing is quite satisfactory and will produce a clean copy from a dirty original if the background dirt is read as a white level. A printer capable of reproducing grey tones (sometimes referred to as half tones) is only necessary if true photographic or picture reproduction is required. The relation between the light from the pixel and the transmitted signal is shown in *Figure 7.2*, and to produce the grey tones the transmitted signal must either be analogue, or contain more than 1 bit per pixel.

Equipment available for connection to the PSTN lines conforms to the ITU-T recommendations for groups 1, 2 and 3 facsimile, summarized in *Table 7.1*.

66

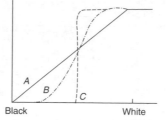

Figure 7.2 Facsimile light characteristic: *A*, linear for accurate photographic reproduction of grey tones; *B*, high contrast – quick transition from black to white; *C*, digital – white or black only

A further group (group 4) is available for high-speed 64 kbit/s circuits, but is only suitable for use on private circuits, ISDN networks and LANs capable of accepting 64 kbit/s digital signals. For such networks, however, the group 4 standard provides very high transmission speeds, an A4 copy taking only 2–4 seconds.

Of the groups 1, 2 and 3, 1 is now obsolete and 2 becoming obsolete. Group 3 is faster (partly because it provides for skipping white areas), provides better reproduction and is digital, and therefore interfaces with other digital equipment when required, or with the PSTN via a modem. The system of modulation employed in the modem, MPSK, is similar to QPSK but employs more increments of phase shift.

The modems used in Group 3 facsimile machines have progressively increased in speed capability. The original Group 3 machines used V.27 ter standard modems that operated at 2.4 or 4.8 kbit/s. Group 3 machines produced in the early 1990s used V.29 standard modems which could operate at 7.2 or 9.6 kbit/s. During the mid-1990s V.17 standard modems were used to raise the data rate to 12 or 14.4 kbit/s. In the late 1990s, V.34 modems operating at speeds up to 28.8 kbit/s were introduced.

Group 4 facsimile machines interface to ISDN terminals, but have the capability to work to Group 3 machines.

The printer operates in a similar manner to the scanner, and usually employs the same mechanical parts. There are a number of alternative methods for printing, the choice depending on the type of scanning employed. These are:

(a) Ink jet, using a controlled jet of atomized ink. This again requires mechanical horizontal scanning. It is also difficult to control for grey tone reproduction.

Table 7.1

ITU-T group	Horizontal resolution (pixels/mm)	Vertical resolution (pixels/mm)	Grey tone capability	Modulation (modem)	Handshake (T.30)	Picture compression	A4 page time (min)
1 (1971)	4	3.85	Option	FM (black 2100 Hz, white 1300 Hz)	Tone	None	6
2 (1976)	4	3.85	Option	AM (SSB) carrier 2100 Hz lower SB	Tone FSK option	None	3
3 (1980)	8	3.85 (option 7.7)	Option, at reduced speed	MPSK, V.27 ter at 2.4/4.8 kbit/s, QAM option V.29 at 7.2/9.6 kbit/s, option V.17 at 12/14.4 kbit/s, option V.34 at 28.8 kbit/s	FSK (V.21) 300 bit/s, 1650 Hz and 1850 Hz	Horizontal, run length, option vertical	<1
4 (1984)	7.87, option 9.45, option 11.8, option 15.7	7.87, option 9.45, option 11.8, option 15.7	Option, at reduced speed	ISDN, 64 kbit/s	7 messages in each direction before data transmitted (4 seconds)	2 dimensional, modified modified read (MMR)	<0.25

(b) Thermal printing where the waxed surface of the paper is burnt away to form the print. This can be achieved by small resistive heating elements each printing one pixel. These are arranged in an array along the horizontal line and energized in groups, allowing each element sufficient time to heat and print. The system eliminates the need for mechanical horizontal scanning, and is therefore a popular printing technique with group 3 facsimile. A disadvantage is the need to use special paper for printing.

(c) Laser printing, where the laser beam is scanned across a selenium coated drum. The drum becomes electrostatically charged at the points which the laser beam has illuminated. The charged surface attracts carbon powder which is transferred to paper as the drum rotates. The paper is then heated so that adhesive attached to the carbon powder melts and fixes the carbon to the paper surface, thus creating a readable image. Laser printing is the most popular for facsimile machines produced after the mid-1990s. One reason is because normal paper can be used, but another reason is that the cost of laser printers has fallen because of their use in many computer systems.

Facsimile is more tolerant of bit errors than telex, since an error will affect only one pixel, which may distort a character but not change it. Parity bits are not therefore necessary with facsimile digital transmission. Handshake signalling is required to establish compatibility between send and receive equipment before transmission commences.

7.1 Data compression

Group 3 and Group 4 facsimile are much faster than Groups 1 and 2 due to the use of data compression. Data compression can be applied because of prior knowledge about the data source; it is a sheet of paper which is largely white, but with black markings. Thus sending individual picture element (pel) information would be pointless if there are several lines of blank paper. The amount of data can be reduced if the number of contiguous white pels are counted, and then a binary code transmitted to represent that number. Similarly, the number of contiguous black pels are counted, and a binary code representing that number is transmitted.

The code words to represent the number of white and the number of black pels has to be designed carefully, so that the number of bits transmitted is minimized. The pattern of the codeword has to be unique so that it can be recognized at the receiving end, without being misinterpreted as another code. To minimize the amount of data, the code words have differing lengths, with the most common black or white pel run lengths having the shortest code. This is similar to the Morse code where the character E is represented by a short tone burst (a dot), while the letter O is three long tone bursts (dashes). The letter E is the most common letter in the alphabet, so it is assigned the shortest code.

Group 3 facsimile uses Modified Huffman Code (MHC). Each scan line contains a minimum of 1728 pels. The coded data always begins with the number of white pels; this may be zero if the first pel is black. If the number of contiguous black or white pels is 63 or less, a terminating codeword (up to 12 bits long) is sent. If the run of single colour pels is longer than 63, the codeword is made up of a make-up code followed by a terminating code. There are individual terminating codewords for all pel numbers from 0 to 63. Make-up codewords are available for multiples of 64 pels up to 1728 pels. Suppose a black run of 153 pels is detected – a make-up code for 128 pels (000011001000) and a terminating code for the remaining 25 pels (00000011000) are used. Thus the code 00001100100000000011000 is transmitted.

Each line ends with an end of line (EOL) code, which is 11 zeroes followed by a one (000000000001). If the receiving machine requires time to restore the printing mechanism, it exchanges data to that effect during the initial handshaking messages. The transmitting machine will insert zeroes in the transmitted data, prior to the EOL message, to create a pause.

Examples of the terminating codes are given in *Table 7.2*. Some make-up codes are given in *Table 7.3*.

Group 3 facsimile has the option of using modified READ (MR) coding. This uses the fact that adjacent lines are likely to have black and white pels in similar positions, which is generally true for text. It is certainly true for the white lines between text. The rules allow for runs of black or white pels where the offset from the line above is no more than three pels. Where the offset is greater than three pels, the modified Huffman code is sent instead. The advantage of modified READ is that the codewords are between 1 bit and 7 bits long, and are only sent when there is a change of colour. An EOL codeword is sent at the end of each line, as before. The MR coding scheme allows for transmission errors by sending MHC typically every four lines. A table of codewords for Modified READ is given in *Table 7.4*.

Table 7.2 Terminating codes

Run length	White codeword	Black codeword
0	00110101	0000110111
1	000111	010
2	0111	11
3	1000	10
4	1011	011
5	1100	0011
6	1110	0010
7	1111	00011
8	10011	000101
9	10100	000100
...		
62	00110011	000001100110
63	00110100	000001100111

Table 7.3 Make-up codes

Run length	White codeword	Black codeword
64	11011	0000001111
128	10010	000011001000
192	010111	000011001001
256	0110111	000001011011
...		
1728	010011011	0000001100101
EOL	000000000001	000000000001

Table 7.4 Modified READ code list

Pel 'a' relative to pel 'b'	Code
Underneath	1
1 pel right	011
2 pel right	000011
3 pel right	0000011
1 pel left	010
2 pel left	000010
3 pel left	0000010

An illustration of modified READ coding is given in *Figure 7.3*. In this case a0 is one place to the right of b0, so the code 011 is sent. The white pels starting at a1 are two places to the right of b1, so the code 000011 is sent.

Group 4 facsimile uses modified modified READ (MMR) coding. This uses the fact that ISDN transmission is very reliable, so no

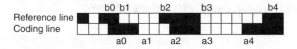

Figure 7.3 Modified READ coding

EOL codewords are sent. Also, MHC does not have to be sent every four lines. These changes allow the amount of data to be reduced considerably and allow a much faster page transmission rate.

8 High-speed data systems

Communication between computers may be by cable (e.g. leased line or PSTN), by HF radio directly between terrestrial stations or by microwave satellite link. A business user may be connected via a combination of these to an associate; for example, a simple small computer terminal may be linked via a telephone line to a local centre communications computer which concentrates its data with that of other local users and passes it via a national centre to a satellite station. From there it could be transmitted by satellite radio link to anywhere in the world, to be reconverted to a data train sent by land line to the ultimate recipient.

8.1 Local data transmission

Data within a computer or between a computer and its peripherals is normally exchanged in parallel format, for example eight data bits, forming a character, and sent simultaneously via eight data wires in a cable from computer to printer. A coding system converts alphabetical and numerical characters into the binary form which a computer requires.

A full 36 character alphanumeric set can be provided by a binary-coded decimal system (BCD) in which the first 4 bits, called the basic quartet, are used four times over in combination with two further bits called the control duet:

Control duet		
0	0	Letters A–I (9 letters)
0	1	Letters J–R (9 letters)
1	0	Letters S–Z (8 letters)
1	1	Numbers 0–9 (10 digits)

Data communications requires an extended character set to include punctuation marks and control characters. The latter perform such tasks as transmission start/stop, formatting, etc.

The extended binary-coded decimal interchange code (EBCDIC) uses 8 bits instead of 6 and out of 256 possible combinations currently has 109 assigned characters. It is commonly used in IBM machines.

The American Standard Code for Information Interchange (ASCII) is widely used and is based on an 8 bit code consisting of seven data information bits plus one error-checking parity bit. The ASCII code uses the International Alphabet No. 5 for control characters which are:

(a) *Transmission controls*

ACK	affirmative from receiver to sender
NAK	negative from receiver to sender
SOH	first character in a heading of an information block
STX	start of text/end of heading
ETX	end of text
ETB	end of transmission block (e.g. of data)
EOT	end of transmission
ENQ	enquiry (e.g. for remote station identity)
SYN	synchronous idle character to maintain sync
DLE	data link escape; changes the meaning of a limited number of succeeding characters (for supplementary control purposes)

(b) *Formatting*

BS	back space
CR	carriage return
FF	form feed
HT	horizontal tabulation
VT	vertical tabulation
LF	line feed

(c) *Separators*

FS	file separator
GS	group separator
RS	record separator
US	unit separator
DC1, DC2,	device controls to switch terminal
DC3, DC4	equipment

(d) *Miscellaneous*

CAN	cancels preceding data
DEL	obliterates unwanted characters
EM	end of medium (e.g. end of tape)
ESC	similar to DLE
SI	shift in – following code combinations are to standard character set
SO	shift out – following code combinations before the next SI are of different meaning to standard text
NUL	blank tape – used for space fill where there is no data

The complete International Alphabet No. 5 and the relation between ASCII characters and the 8 bit binary code is given in Chapter 26.

Note that the tabulation is also in ascending decimal order from 0 to 127.

When the sending distance becomes more than a few metres, conversion of the parallel format binary code to a series format is necessary so that a two-wire (or four-wire full duplex) system can be adopted. The conversion is performed at a parallel/series interface, the simplest form of which is a clocked shift register. This (for an 8 bit character) consists of a cascade connection of eight flip-flop (bistable) circuits, which are all cleared before each character is entered. Parallel entry of the 8 bits applies a '1' to the preset input of each stage which is to be set, the remaining stages staying in the reset condition. The isolated data is read out of the register, one bit at a time, under control of the synchronizing clock. At the receiving station, the series data is converted back to 8 bit parallel format by another shift register connected for series input/parallel output. In practice a buffered system with an intermediate holding register is used and this permits the computer to check and transfer a character from the buffer while the next character fills the input register.

LSI circuits are available to combine the transmit and receive functions; the USART (Universal Synchronous/Asynchronous receiver/transmitter) is made in versions to support 4, 8, 16, 32, 64, 128 and even 256 parallel input/output lines. Asynchronous operation, in which information is transmitted in packets, each packet containing the necessary data to decode the information it contains, is the more common. It is more complex than a synchronous system, but a separate clock signal is not required.

In principle the receivers and transmitters use a scanning system so that, for example, when a received character is located it is transferred to a first-in-first-out (FIFO) buffer store along with its line identification. The central processor unit (CPU) is notified of a character in FIFO and accepts it. Similarly when an empty transmitter buffer is scanned the CPU is notified and sends the next character to that location. Serial data transmission with the TTL-compatible signals from a UART (or USART) is only possible over very short distances (e.g. a few metres). Before transmission over any appreciable distance (e.g. from one part of a building to another) a line driver is required to provide buffering and level shifting between the UART and transmitting system.

Likewise a line receiver is required to reconstitute the incoming signal to TTL levels. The transmission signal will generally be a two-level one such as ± 6 volts thereby avoiding an error-prone zero voltage.

8.2 56 k modems (V.90 and V.92)

The so-called 56 k modems were not standardized initially, because the two companies who developed the technology kept the development secret from everyone until the product was ready for marketing. Rockwell produced a modem called 56 k, whilst US Robotics produced a modem called X2. The two designs are now changing to a common ITU-T V.90 standard, being implemented during 1998.

The V.90 modem was developed specifically for Internet access. Unlike conventional modems, the V.90 modem relies on the Internet Service Provider (ISP) having access to a digital connection in the telephone system. The digital data is transmitted from the ISP through its local exchange and onto the Internet subscriber's local exchange. The digital code is then converted into analogue signals. These signals are actually voltage levels representing the digital code, and each voltage level can represent up to 8 bits of data, as shown in *Figure 8.1*. Since 8 bits of data are transmitted every 125 microseconds (8 kHz sampling rate), a maximum of 64 kbit/s can be transmitted in theory.

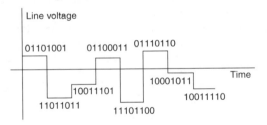

Figure 8.1 V.90 line signals

This method of signalling cannot achieve 64 kbit/s in practice. The digital to analogue converter (DAC) in the Internet subscriber's local exchange does not give a linear output. Instead it uses A-law (E. 1) or μ-law (T.1) encoding which graduates the output voltage to give small voltage steps when close to zero output. This encoding is used to give a constant signal to noise output for a wide range of analogue voltage levels, as illustrated in *Figure 8.2*. Unfortunately, for V.90 applications, it is difficult for the receiver to discriminate between the small voltage steps for codes represented by voltages close to zero. Consequently, the data rate is limited to 56 kbit/s or less.

In conventional modem links, a modulated tone is transmitted from the ISP into the ISP's local exchange. Here, it is digitally encoded and transmitted to the Internet subscriber's local exchange before being

76

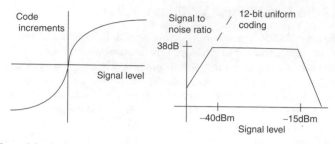

Figure 8.2 Analogue to digital coding

converted back into an analogue tone for transmission over twisted pair to the end user.

Transmission of data in the reverse direction is identical for conventional and V.90 modem applications. The V.90 modem uses stepped voltage levels for transmission to the Internet subscriber, but conventional modem signals (i.e. a modulated tone) for transmission back to the ISP. The modulated tone transmitted by the Internet subscriber is received in their local exchange and converted into a digital signal for transmission to the ISP's local exchange. Here, it is converted back into a tone and transmitted to the ISP over twisted pair. The ISP use a conventional to demodulate the subscriber's data.

The comparison between conventional and V.90 modem transmission is illustrated in *Figure 8.3*.

The V.92 modem allows up to 56 kbps in the downstream path, but raises the data rate in the upstream direction from 33.6 kbps to 44 kbps.

8.3 High-speed digital subscriber line (HDSL)

The HDSL system was developed to provide a means of extending 2 Mbit/s E.1 or 1.544 Mbit/s T.1 data to subscribers when the only access cable is twisted pair. A typical application is the provision of multiple telephone circuits to private branch exchanges (PBXs) or high speed local area network (LAN) connections. Current systems require two or three pairs; the data is shared between the pairs so that the data rate on each pair is reduced, and this allows the line length to be up to 3 miles (5 km).

The data on each pair is encoded by the 2B1Q method. Two binary bits (2B), or dibits, are converted into one of four voltage levels (1Q). The DC voltage levels are transmitted along each pair and measured at the receiver for decoding back into dibits. This is fairly

77

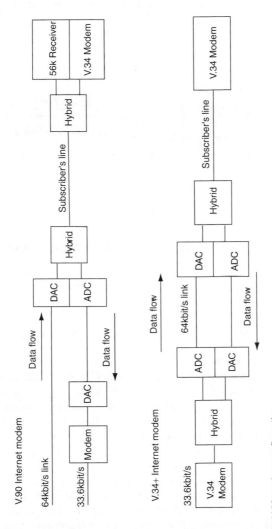

Figure 8.3 V.90 modem configuration

straightforward, except that data is simultaneously being transmitted in the opposite direction. Echo cancellation techniques are used to subtract the transmitted signal from the composite signal across the line, thus leaving the received signal. A sample of 2B1Q data is shown in *Figure 8.4*.

The ITU-T standard for HDSL is G.991.1.

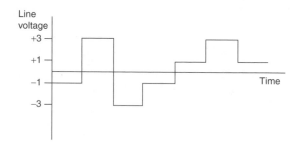

Figure 8.4 2B1Q data

8.4 Asymmetric digital subscriber line (ADSL)

The ADSL system was developed as a result of competition from cable television companies. Cable TV companies provide telephone services alongside their television service, often at low-cost, as a marketing tool to sell their product. The ADSL system allows incumbent telephone operators to use existing twisted pair cables to provide television services; the aim is to reduce the number of customers wishing to change service provider. The system is asymmetric because the data rate from the exchange to the subscriber is much greater than in the opposite direction. A data rate of 2 Mbit/s is required in order to transmit compressed video signals. Such an asymmetric system would also be suitable for Internet access, where far more data flows from the ISP than from the Internet subscriber.

The ADSL system does not use 2B1Q line coding. The design criteria for ADSL is that it should use the existing telephone service pair, thus the telephone must operate as before using a DC loop to indicate an active line. Since the 2B1Q line coding uses DC voltage levels, it is incompatible with DC loop conditions. The ADSL system uses the spectrum above audio frequencies for transmission, as shown in *Figure 8.5*. Basically, the ADSL system is a wideband modem.

There are two modulation methods. One is called discrete multi-tone (DMT) uses a number of discrete carrier frequencies, each

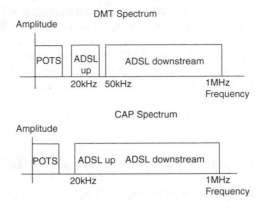

Figure 8.5

one being modulated by part of the data. The modulation spectrum of each carrier occupies a 4 kHz bandwidth. This is like having lots of modems running in parallel, each with a different carrier frequency. The second modulation method is called carrierless amplitude and phase (CAP) and is similar to quadrature amplitude modulation (QAM). Instead of having a carrier that is amplitude and phase modulated, the CAP system synthesizes an equivalent waveform.

Line modulation using DMT is specified in the ANSI T1.413 standard, and is co-specified with CAP modulation in ITU-T Specification G.992.1.

The disadvantage of using high frequencies for transmission is that line attenuation increases with frequency. This means that those signals which use the higher frequencies are attenuated more than those using frequencies just above audio and the working range is limited. The advantage that ADSL has over HDSL is that only a single pair is required. The disadvantage is that because of line attenuation considerations, high level signals are transmitted in the high data rate direction. To prevent near-end crosstalk problems, the high data rate direction must always be from the exchange to the customer.

8.4.1 ADSL (G.lite)

A low-cost version of ADSL avoids the need for a filter in the customer premises to separate the POTS and ADSL signals. It is known as splitter-less ADSL and is defined by ITU recommendation G.299.2.

The line modulation uses DMT and allows a data rate up to 1.5 Mbps in the downstream direction towards the subscriber and up to 512 kbps in the upstream direction towards the service provider.

Conventional ADSL using DMT modulation occupies 256 frequency bands, each 4 kHz wide. These bands start at 25 kHz and extend up to 1.049 MHz. The alternative G.lite uses 96 frequency bands and therefore its line borne signal has less high frequency content than standard ADSL, with a maximum frequency of 409 kHz.

8.5 Single-pair digital subscriber line (SDSL)

Symmetrical digital subscriber line working over a single pair, at variable data rates up to 1.1 Mbps, is possible using SDSL. The line modulation can be either 2B1Q or CAP. The SDSL system is likely to be made obsolete in the future by the use of H.shdsl.

8.6 High-speed digital subscriber line 2 (HDSL2)

Operation of HDSL over a single pair is possible using HDSL2. This is intended to carry fixed rate T1 (1.544 Mbps) or E1 (2.048 Mbps) signals, like HDSL. A 16-level PAM line code is used, with spectral shaping. The symbol rate transmitted over the line is 517.3 k or 682 k symbols/s. Each symbol represents four bits: three data bits and one forward error correction (FEC) bit. The FEC process uses a single-stage 512-state Trellis-coded modulation. The power spectral density of the modulated signal is not flat, but uses OPTIS shaping. The HDSL2 technique is standardized by the ITU-T as G.991.2.

8.7 Single-pair HDSL (G.shdsl)

Although not yet standardized, G.shdsl is intended to provide symmetrical data transmission at rates of up to 2.3 Mbps. Variable data rates will be possible, from 144 kbps to 2.3 Mbps. Like the HDSL2 system, a 16-level PAM line code is used. However, the G.hsdsl system does not apply spectral shaping to the signal before it is transmitted: the power spectral density is flat. This allows the maximum line length to be 20% greater than that possible for HDSL2.

8.8 Very-high-speed digital subscriber line (VDSL)

Very high speed digital subscriber line (VDSL) systems are not yet standardized. The purpose of VDSL is to provide a very high data rate into offices and buildings. The system will be asymmetric, in the early models at least, allowing a maximum upstream rate from the customer to the exchange of about 2 Mbit/s. The downstream data rate is expected to be up to 55.2 Mbit/s, which limits the range over twisted pair cables to about 300 m. Lower data rate systems permit a longer range over twisted pair cable. Limiting the downstream data rate to 27.6 Mbit/s will allow a range of 1 km. Further limiting the rate to 13.8 Mbit/s allows a working range of 1.5 km.

Transmission of data to outside the building will use optical fibre as a 'fibre to the kerb' (FTTK) system. VDSL will perform the final drop into the customer's premises.

8.9 The integrated services digital network (ISDN)

The preamble to the relevant ITU-T recommendations gives the following general description:

An ISDN is a network, in general evolving from a telephony integrated digital network (IDN) that provides end-to-end digital connectivity to support a wide range of services, to which users have access by a limited set of standard multi-purpose user network interfaces.

First discussions and proposals date back to the early 1970s and a considerable amount of work has been expended into standardization, interfacing and field trials on a limited national basis. Today, PSTNs are still a mix of analogue and digital transmission bearers providing public service, private line networks and both circuit and packet switched data networks. Subscriber equipment includes telephones, facsimile, Telex, data terminals, personal and mainframe computers. *Table 8.1* lists the basic telecommunications services provided by an ISDN.

It is essentially improvements in technology, and in particular the common-channel signalling protocol offered by ITU-T SS No. 7 (Known as C7), that are making the realization of the initial ISDN conception a practical reality for the 1990s although many problems of standardization remain in the international field.

Table 8.1

Bearer services (i.e. transparent information channels between terminals):

Circuit mode speech
Circuit mode 64 kbit/s unrestricted
Circuit mode 3.1 kHz audio
Circuit mode 2 × 64 kbit/s unrestricted
Packet mode X.31 case A (B channel)
Packet mode X.31 case B (B channel or D channel)

Teleservices (i.e. user-to-user services):

Telephony (3.1 kHz bandwidth)
Teletext
Group 4 facsimile
Telephony (7 kHz bandwidth)
Audiographic conferencing
Teleaction
Videotelephony
Computerized communication service

A wide range of supplementary services (e.g. charging information, diversion services, freephone, etc.) is also available.

Figure 8.6 shows the basic difference between the existing network structure and an ISDN. The economic and operational advantages of universal common connectivity are obvious.

8.9.1 The ISDN user interface

Figures 8.7–8.9 show three different configurations of user equipment offering single access, multiple access and multiple-access passive bus. Note the difference between *Figure 8.8* and *8.9* is the inclusion/omission of NT2. Reference points need not exist as a physical interface, since two functional blocks may be combined within one piece of equipment. The reference points and functions on the diagrams are described as follows:

1 R reference, between a non-ISDN terminal and its terminal adaptor, will be a non-ISDN interface such as X21, V35, RS232C.
2 S reference is between an ISDN terminal (or terminal adaptor) and PBX, LAN or other non-ISDN terminal equipment. A 'fan-out' of up to eight ISDN terminals (or adaptors) is permitted via the S interface from a PBX. The S reference is a two-pair four-wire link via an eight-way connector.
3 T reference is a point on the two-pair four-wire interface between ISDN terminal equipment (or PBX or terminal adaptor) and the network termination.

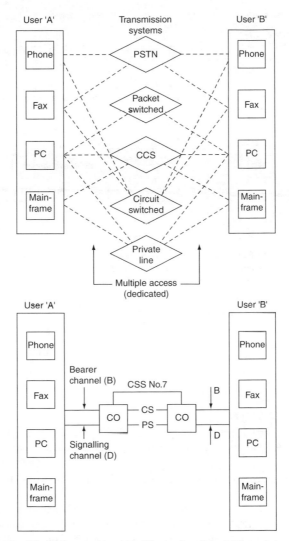

Figure 8.6 CO, ISDN central terminal; CS, circuit switched; PS, packet switched; CCS, common-channel signalling system

4 U reference is the transmission line between the ISDN exchange and the customer network interface. This is a full duplex interface over a twisted pair for basic rate (64 kbit/s), but over two pairs for primary rate (1.5 or 2 Mbit/s). International standardization of the

84

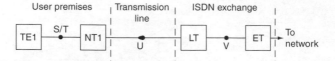

Figure 8.7 Single-user access (point to point): TE1, ISDN-compatible terminal equipment; NT1, network termination type 1; LT, line termination; ET, exchange termination; -•-, reference points

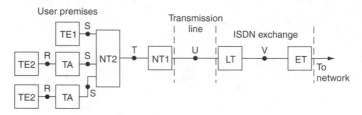

Figure 8.8 Multiple-user access (point to point): NT2, network termination type 2; TA, terminal adaptor; TE2, non-ISDN-compatible terminal equipment. For other symbols see *Figure 8.7* caption

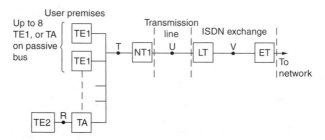

Figure 8.9 Multiple-user, passive bus access at base rate only (point to multipoint)

U reference point (e.g. in respect of echo cancellation) is still to be finalized.

5 V reference is between the line termination and exchange termination within the ISDN exchange. If these two are the same equipment the V reference, which is not yet standardized, would not exist.

The functional blocks either side of these reference points can be subdivided into network terminations which deal with communications on the network side of the customer/network interface, and terminal equipment which includes telephones, data terminals, etc. providing communications on the user side:

6 Network termination 1 (NT1) provides signal conversion, timing, and physical and electrical termination of the network at the customer's premises. It corresponds to layer 1 of the OSI architecture.

7 Network termination 2 (NT2) corresponds to OSI layer 2 (data link) and often also layer 3 (network) functions. Typical physical realizations are front-end processors (FEPs), concentrators, multiplexers, PBX and LAN.

8 Terminal equipment 1 (TE1) has an ISDN-compatible customer/network interface and would be, for example, a digital telephone, personal computer or data terminal. It provides protocol, interface and connection functions to other equipment.

9 Terminal equipment 2 (TE2) does not have an ISDN-compatible interface but is otherwise functionally as a TE1. Typical TE2s would be personal computers or data terminals with an RS232, V35 or X21 interface.

A TE2 requires a terminal adaptor (TA) to convert non-ISDN message protocols and bit rates to ISDN standards.

The local cabling installed to satisfy telephone and low-speed data requirements, based on an audio band of 300–3400 Hz, represents a very high capital investment. The most advantageous way of adapting this cable network to ISDN requirements as detailed above has been, and still is, the subject of extensive investigation. The ISDN 2B + D requires a bit rate of 144 kbit/s, which as an unprocessed baseband signal requires a transmission band of at least 72 kHz. Attempts to preserve the existing cables as duplex two-wire connections for each subscriber result in problems from high-frequency attenuation and cross-talk. Initial investigations considered 'burst mode' operation, storing and transmitting (at a higher bit rate) go and return alternating to provide full duplex operation on each pair. This was also referred to as 'ping-pong'.

Digital processing using digital signal processors (DSP) has developed to the extent that it is now economic to consider quite complex processing in the subscriber to exchange access. It is 2B1Q line coding, with echo cancellation, which enables an increase in bit rate without an increase in the transmitted baud rate. This is now common in the UK, Germany and the USA for the basic rate ISDN connection. HDSL (Section 8.3) uses these techniques on two or three pairs to provide 30 64 kbit/s channels. This would enable the existing local line network to be rearranged for four-wire working (with echo cancellation if required), and at the same time release pairs for further expansion of the service. It also makes possible the combining of channels to provide faster bit rate services to local subscribers (e.g. video conferencing requiring six channels, 384 kbit/s).

It would appear therefore that not only will local digital access be achieved using most of the existing distribution cable network, but the cables will be used more efficiently, permitting expansion without the need to provide additional cable capacity. Additional capacity when needed, will probably be provided by fibre optic cables or local access radio.

8.9.2 The basic rate interface (BRI)

In the ISDN information is transferred between network terminations and terminal equipment by specified channels of which the most widely adopted are the so-called B and D channels.

The B or bearer channel is a 64 kbit/s voice-grade digital channel carrying voice and data transmissions in circuit switched or packet switched mode. It does not carry signalling information used to set up and control the call, but it is bit transparent for data transmission (including voice, text and video).

The D or signalling channel is a 16 kbit/s (or sometimes 64 kbit/s) channel dedicated principally to signalling information (as in ITU-T SS7). The utilization of the D channel for signalling normally does not approach 100%, since signalling information occurs in relatively short bursts. The spare capacity may be used for packet switched customer data.

The basic rate interface consists of two 64 kbit/s B channels plus one 16 kbit/s D channel and so is often referred to as a 2B + D interface. The data handling capacity is $(2 \times 64) + 16$, that is 144 kbit/s, but additional control bits may be added to raise the total rate to 192 kbit/s. Since the two B channels are totally independent they can be used, for example, for simultaneous transmission of voice and data over the same transmission link.

8.9.3 The primary rate interface (PRI)

This is exactly analogous to the primary PCM multiplex and is formed by time division multiplex in just the same way. In Europe 30 B channels, one D channel, plus one overhead (control) channel each at 64 kbit/s are multiplexed to give a total rate of 2048 kbit/s, whilst in the USA, South Korea and Japan 23 B channels and one D channel at 64 kbit/s plus an overhead at 8 kbit/s are multiplexed to 1544 kbit/s. In the European system, as for PCM telephony each 125 μs frame carries 32 time slots containing 8 bits. TS0 is used for frame alignment,

TS16 carries the D channels of all 30 B channels and TS1–15, TS17–31 carry the 30 B channels. In the 1544 kbit/s system, each 125 μs frame has 193 bits, one of which is used for framing; the remaining 192 bits are divided into 24×8 bit time slots, 23 of which are for the B channels and one for the D channel. The composite D channel* in both cases carries the signalling for all of the associated B channels and is at 64 kbit/s.

Both BRI and PRI may be present at reference points S, T or U.

8.9.4 Protocols

All of the information necessary to establish, supervise and clear down an ISDN connection is carried by the D channel. A three-layer protocol (based on an OSI model) is used.

Layer 3, the network layer which provides the means to establish, maintain and clear down network connections across an ISDN exchange, uses a similar message structure to the X25 packet layer. The layer 3 message structure is shown in *Figure 8.10*.

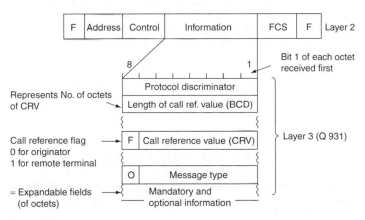

Figure 8.10

This format is based on ITU-T Q931 and controls call establishment, flow control and error correction, call cleardown and miscellaneous service/maintenance information.

Layer 2 ensures error-free data transmission across the user/network interface and is controlled by a local access protocol (LAP-D) which is defined in ITU-T Q921.

---

* Note: The term 'D' channel has been found to be used for both the 16 kbit/s BRI and the 64 kbit/s at a PRI.

88

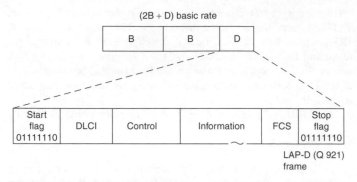

Figure 8.11 DLCI, 16 bit address (data linked control identifier); control, 8 or 16 bit; information, actual message packet for layer 3 (up to 256 bytes); FCS, frame check sequence (error check)

Its basic frame structure is shown in *Figure 8.11*.

Layer 1 protocols cover the initial handshaking between terminal equipment and network termination within the customer's premises. The message format is much simpler and consists of four specific signals called INFO signals which are defined in ITU-T I.430. The four signals (other than INFO '0' which is no signal and is all binary 1s) are:

INFO 1 Sent from a TE to activate an NT, when the TE is both sending and receiving all binary 1s (i.e. INFO 0).
INFO 2 Sent by an NT for one of four reasons:
    (a) to activate a TE once it is sending INFO 0;
    (b) if it receives INFO 1 once it is sending INFO 0;
    (c) if it receives INFO 0 once it is sending INFO 4;
    (d) if it loses frame sync whilst sending INFO 4.
INFO 3 Sent from TE to NT when frame alignment is established.
INFO 4 Sent by an NT if it receives INFO 3 once it is sending INFO 2.

INFO 0 is sent by a TE on 'power on' loss of frame alignment or if it receives any signal once it is sending INFO 1.

INFO 0 is sent by an NT to deactivate or clear down a TE once the NT is sending INFO 2 or INFO 4.

8.9.5 ITU-T and ISDN

The I series of ITU-T recommendations were agreed after a 4 year study period in 1984. They cover the user network interface, service aspects, network aspects and matters of general ISDN interest, but exclude ISDN maintenance principles or internetwork interfaces, recommendations on which may be expected shortly.

The I series has six separate parts of which the I–100 group forms the central core covering the concept, principles and vocabulary of the ISDN. The six sections cover in summary:

I-100 series – the general concept, structure and terminology
I-200 series – service capabilities
I-300 series – overall network aspects and functions
I-400 series – user/network interface
I-500 series – internetwork interfacing
I-600 series – maintenance principles

Those who may wish to consider the benefits of ISDN working when it becomes available to them would be well advised to refer at least to the I–400 series which defines the all-important interface between user and network so as to have some first-hand knowledge of internationally agreed recommendations.

8.9.6 Line codes for basic rate ISDN

The ITU has agreed a dual standard for ISDN basic rate line codes, see G.961. One ISDN line code standard is widely used outside Europe and is described by ANSI specification T1.601. This uses 2B1Q line coding, which was described in Section 8.3 (see Figure 8.4). Data is sampled as a group of two binary bits and these are coded into one of four line voltages, see *Table 8.2*.

Table 8.2 2B1Q line code mapping

Data	Line symbol	Voltage
00	−3	−2.5 V
01	−1	−0.833 V
10	+3	+2.5 V
11	+1	+0.833 V

In Europe, a 4B3T line code is used for ISDN. Data is sampled as a group of four binary bits, which gives 16 possible combinations for each group. A group of three symbols is transmitted to line for every four data bits sampled. Each symbol has one of three voltage levels: positive, negative or zero volts. For example, a data group of

0010 is coded and transmitted to line as 0−+. Note that summing these symbols produces zero volts. An average line voltage of zero is important because it allows lines to be a.c. coupled, usually by using a transformer.

Three symbols, each having three possible amplitudes, gives 27 possible combinations. Only seven combinations give zero volts when the symbols are summed. Of these, the combination 000 cannot be used because this group of symbols contains no timing information. So, six groups of three symbols can be directly matched to six groups of four data bits.

The remaining 20 combinations of line code, which produce either a positive or negative voltage when summed, are matched to the remaining 10 combinations of data groups. This ratio of 2:1 is fortunate because it means that two line code combinations can be allocated to each combination of data. Choosing the two line codes that have the opposite polarity in each symbol (e.g. ++0 and −−0) allows decoding to be straightforward.

The transmitter sums the voltage from every symbol transmitted to line. This is known as the Running Digital Sum (RDS). Where the transmitter has a choice of two line codes, it chooses the one that raises or lowers the RDS, as appropriate. Its aim is to keep the RDS at, or near, zero.

The mapping between data and line codes is given in *Table 8.3*.

Table 8.3 4B3T line code mapping

Data	Line Code 'A'	Line Code 'B'	Sum
0000	+0−	+0−	0
0001	−+0	−+0	0
0010	0−+	0−+	0
0011	+−0	+−0	0
0100	++0	−−0	+/−2
0101	0++	0−−	+/−2
0110	+0+	−0−	+/−2
0111	+ + +	− − −	+/−3
1000	+ + −	− − +	+/−1
1001	− + +	+ + −	+/−1
1010	+ − +	− + −	+/−1
1011	+00	−00	+/−1
1100	0 + 0	0 − 0	+/−1
1101	00+	00−	+/−1
1110	0+−	0−+	0
1111	−0+	+0−	0

8.9.7 Line codes for primary rate ISDN

Primary rate ISDN line codes are HDB3 in Europe and B8ZS in North America. The purpose of these codes is similar to that for the Basic Rate ISDN: to reduce the DC content and maintain timing information.

The HDB3 code is described in Section 21.2. No more than three zeroes can be transmitted consecutively. The B8ZS code allows the transmission of up to seven consecutive zeroes.

9 Copper wire cable

Copper wire has been used for many years as the mainstay of signal transmission. It has low electrical resistance, so the loss due to current flowing in the wires is quite small. It is very easy to work; it can be bent and straightened many times without breaking, i.e. it is malleable.

Copper wire cable used in telecommunications comes in two varieties: (1) twisted pair; and (2) coaxial pair. Each of these will now be discussed in detail.

9.1 Twisted pair cable

A twisted pair is a pair of insulated copper wires, twisted together. The diameter of all the copper wires in a cable are usually the same, although some special cables are available that have a few larger diameter wires in amongst many smaller diameter wires. In these cables, the larger diameter wires are often used for to carry high frequency signals, since the signal loss reduces as the diameter increases.

Typical wire diameters are 0.4 mm, 0.5 mm, 0.63 mm and 0.9 mm. This is the diameter of the copper wire alone. Insulation covers the wires, so the overall diameter of each insulated wire is greater and is typically double that of the wire alone. The insulation material is often PVC or PE, although paper insulation was still used in cables until the 1960s. The type and thickness of the insulation material determines the capacitance between the wires in a pair.

The insulation is colour coded in some way so that the user can identify the wires he wants to use. The older paper insulated wires had coloured bands at intervals along the wire. Modern plastic coated wires are colour coded; often in a standard format. Each wire has two colours: a primary colour and a secondary colour. The primary colour covers most of the wire. The secondary colour is either a lengthways stripe or bands at intervals around the wire. A table of colour codes is given in Chapter 26.

Modern cables have an overall PVC or PE sheath. Internal cable sheath is usually PVC and external cable sheath is usually PE. The problem with PVC is that it gives off chlorine gas or other dangerous fumes when burnt. Some internal cables have a special sheath material that does not create dangerous fumes when burnt; this is known as low smoke fume (LSF) cable.

Telecommunication cables can contain as few as two twisted pairs, or as many as 3 000. Generally speaking, cables with large numbers of pairs have a small wire gauge. Two-pair cable with 0.9 mm diameter wire, or 3 000 pair cable with 0.4 mm diameter wire, are both likely to be found. Internal cables usually have 0.5 mm diameter wire, no matter how many pairs they contain (although these cables are limited to about 200 pairs).

In external cables, care must be taken to ensure that water does not come into contact with the telephone wires or joints. Not only will corrosion occur, but the insulation between the two wires will break down; this could cause a number of faults such as: being unable to make calls; false answering on incoming calls; and line noise. To prevent water ingress, cables are either grease filled (with petroleum jelly) or have pressurized air inside them. Pressurized cables rely on an air compressor generating dry air. The air pressure is usually in the range 4–10 pounds per square inch. If the cable sheath is holed, air escapes preventing water from entering.

9.1.1 Twisted pair impedance

The impedance of a twisted pair can be modelled as two resistors in series, with a capacitor in parallel with one of them, as shown in *Figure 9.1*. Diagram (a) shows the capacitive telephony reference impedance, which is valid for audio band frequencies. Diagram (b) shows the high frequency model of a line which is valid for frequencies above the audio band. The line impedance approaches 100 Ω resistive at 100 kHz and beyond.

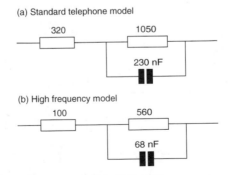

Figure 9.1 Equivalent impedance for a twisted pair

A telephone line does not have an impedance of 600 Ω over the whole of the audio frequency band. The figure of 600 Ω is often used

94

to specify the performance of components such as line isolating transformers; the loss is usually specified between a 600 Ω source and a 600 Ω load. This is to simplify the testing process and give a common benchmark with which to compare similar devices. A telephone line has an impedance of 600 Ω at just one frequency, which is about 1200 Hz for standard 0.5 mm copper wire pairs.

The impedance of a twisted pair is given by the equation:

$$Z_0 = \sqrt{\frac{R + j\omega L}{G + j\omega C}}$$

At high frequencies the impedance becomes:

$$Z_0 = \sqrt{\frac{j\omega L}{j\omega C}} = \sqrt{\frac{L}{C}}, \text{ since the } j\omega \text{ terms cancel.}$$

Note that this is purely resistive.

The impedance of a copper pair using 0.5 mm diameter wire is graphically illustrated in *Figure 9.2*. The frequency axis is in logarithmic steps, to show the detail over a wide frequency range. The frequency doubles at each step: for illustration, the impedance at 300 Hz, 600 Hz, 1.2 kHz, 2.4 kHz, 4.8 kHz, etc., has been measured. The impedance is about 1500 Ω at 300 Hz, falling to 100 Ω at frequencies above 100 kHz.

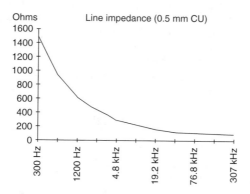

Figure 9.2 Impedance of copper pair using 0.5 mm wire

9.1.2 Twisted pair signal attenuation

Attenuation of signals in twisted pair cable is very dependent on the signal frequency. The attenuation rises proportional to the frequency, but in the 60 kHz to 120 kHz band the attenuation remains fairly constant. This is shown in the graph in *Figure 9.3* that was created from measurements taken on actual cables. The graph covers the frequency range 300 Hz to 10 MHz using a logarithmic frequency axis.

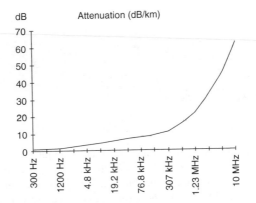

Figure 9.3 Attenuation in twisted pair cables (300 Hz to 10 MHz)

The equations to find the attenuation of cables are:

$$\gamma = \alpha + j\beta$$
$$\gamma = \sqrt{(R + j\omega L)(G + j\omega C)}$$

It is reasonable to assume that $G = 0$, for all modern plastic insulated cables.

If the value of wire inductance is increased artificially, by inserting an inductor in series with the wire, the loss at audio frequencies can be reduced. The ideal condition is when $LG = CR$. However, since G is small this is difficult to achieve. In practice 88 mH inductors are inserted in the line at one mile intervals. These are called loading coils. This practice results in a very low loss at audio, but increases the loss at high frequencies. If inductors are inserted in a line, the line is said to be 'loaded'. This technique is becoming less commonplace now; however in the USA, where long lines are more common, loading coils are still used. The attenuation in a loaded line is illustrated in *Figure 9.4*.

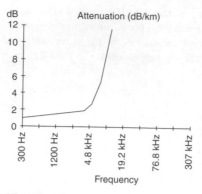

Figure 9.4 Loaded line attenuation

9.1.3 Crosstalk

Crosstalk is when signals on one pair of wires couple into another pair of wires. Clearly crosstalk is undesirable from the privacy point of view. It is also undesirable for modern signalling equipment, which will not work properly if interfering signals are present. Intelligent modems, for example, will reduce their transmission speed if crosstalk from other modems is present.

Crosstalk comes in two varieties: near-end crosstalk and far-end crosstalk. These two varieties are purely for analysis purposes. Near-end crosstalk is the amount of signal into a receiving device from a co-located transmitting device. Far-end crosstalk is the amount of signal into a receiving device when the transmitting device is at the other end of a system. The crosstalk mechanisms are illustrated in *Figure 9.5*.

9.1.4 Transverse screened cable

To reduce crosstalk, particularly near-end crosstalk, transmit and receive pairs are divided into separate halves of the cable. An aluminium foil screen is placed between the transmit and receive pairs. The ends of this foil are wrapped around the outside of the copper pairs to screen them from external fields, so that the foil has an overall S shaped cross-section. This type of cable is called transverse screened cable, and is illustrated in *Figure 9.6*.

The foil screen is earthed at the telephone exchange and provides an electrostatic screen. The levels of crosstalk between the copper pairs is significantly reduced by this technique. Transverse

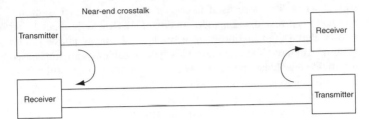

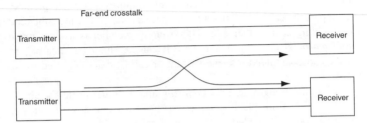

Figure 9.5 Near-end and far-end crosstalk

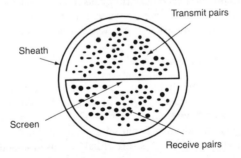

Figure 9.6 Transverse screened cable

screened cable is used to provide high bit rate digital line systems to customers, where separate transmit and receive pairs are used. The type of systems using this cable are N × 64 kbit/s systems such as ISDN30 (2.048 Mbit/s data rate).

9.1.5 External dropwire

Many homes have telephone lines that are still fed by overhead wires; connected to a pole at one end and the house at the other. These wires are known as dropwire, since they are the last drop to the customer from a cable route. The most common dropwire is a single pair of wires. The wire itself is copper covered steel that is strong, resistant to corrosion and a good conductor. Each wire is about 1 mm in diameter. The insulation is most often a polyvinyl chloride (PVC) material, coloured black or grey. This type of cable is sometimes used to provide telephone circuits between a distribution point at the front of a building to customers in adjoining buildings.

Alternatively, multi-pair cables are sometimes used in overhead feeds, more commonly to feed an apartment block or office where several pairs are needed. Overhead cable requires a strength member to support the cable. It is fixed to the pole at one end of the cable and the building at the other end. The strength member is usually made from plastic covered stranded steel. It can be either an integral part of the cable, moulded into the sheath, or can be externally tied to the cable.

Both the strength member covering and the outer cable sheath are usually made from black polythene (PE). All external cable sheaths have carbon black added to the plastic material during manufacture. This is because when polythene (which is naturally translucent) is exposed to sunlight it becomes brittle and can crack easily. Any cracks in the sheath will allow water to enter, causing corrosion or a break-down of the insulation between the wires of a copper pair.

9.1.6 Other copper pair cables

Another copper cable that is quite common is screened twisted pair. Often this cable has just one pair of insulated multi-stranded wires and a tinned copper wire braid covering. Each 'wire' is several strands of thin wire (7 strands of 0.2 mm wire is common). An overall PVC sheath protects the cable. This sort of cable is used in audio systems, typically for loudspeaker connections.

9.2 Coaxial cables

Coaxial cable is familiar to many people for its use in television aerial feeds. This cable has a single inner conductor insulated by solid nylon or polythene. The insulator is then covered by a wire braid. Finally the braid is covered in polyvinyl chloride (PVC). Coaxial cable has the advantage of virtually zero crosstalk and much lower losses at high frequencies compared to twisted pair cable.

Coaxial cable used to connect radio frequency equipment is usually 50 Ω impedance. Cable used for television feeders and telecommunications is usually 75 Ω. Coaxial cable used in telecommunications is of superior quality. Also there are usually several coaxial pairs in each cable. Coaxial pairs are sometimes called tubes, for reasons that will become apparent. The inner conductor is covered in a nylon insulator that is arranged to have low density. One method of achieving this is by 'foaming' the insulator to provide air bubbles. A second method is to form two concentric layers of insulator separated by webs. This is illustrated in *Figure 9.7*.

The outer conductor of a high quality coaxial pair will have both metal foil and braid. The metal foil provides a good screen. The braid maintains electrical connectivity if the foil fractures during installation. To reduce the risk of foil fracture, the cable should not be bent more than necessary. Cable manufacturers may specify a minimum bend radius; if not a few simple experiments can be used to determine a suitable value.

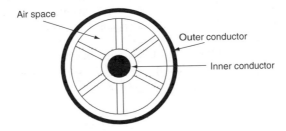

Figure 9.7 Telecommunications coaxial pair (tube)

The size of the 'tubes' can vary considerably. The advantage of using a larger tube is that the inner conductor can have a larger diameter and therefore lower resistance, which leads to lower attenuation. The larger the tube, the lower the attenuation. The disadvantage of using larger tubes is the amount of duct space required and the greater care needed in handling the cable. Larger tubes will have a greater minimum bend radius since this is roughly proportional to the tube diameter. A single tube cable may have a diameter of 15 mm or more. A slightly larger cable of about 22 mm in diameter could have maybe 18 tubes, each about 4 mm in diameter.

9.3 Data transmission cables

A number of screened data transmission cables are available, including twin axial, twisted pair, and multipair. Typical characteristics are:

Twin axial (e.g. Belden 9272, UL 2092):

$Z = 78$ ohm $C = 65$ pF/m diam. 6.5 mm
Attenuation per 100 m: 16.4 dB at 50 MHz 6.9 dB at 10 MHz

Screened twisted pair (e.g. Belden 9279, UL 2493):

$Z = 100$ ohm $C = 41$ pF/m diam. 8.5 mm
Attenuation per 100 m: 6.8 dB at 10 MHz

Multipair (e.g. 10 core Belden 9540):

$Z = 89$ ohm $C = 98$ pF/m core/core; 180 pF/m core/screen diam. 6.4 mm
Attenuation per 100 m: 6 dB at 1 MHz

9.3.1 CAT-3, -4, -5, -6, -7 cable

Twisted pair data cables for LANs (such as 10 base T or 100 base T) are described as category 3, 4, 5, 6 or 7; these are often referred to as CAT-3, CAT-4, etc. CAT-5 is specified by standards TIA/EIA 568A, ISO/IEC 11801, EN 50173. CAT-6 is specified by standards TIA/EIA 568B, ISO/IEC 11801 Category 6, EN 50288. The category determines the maximum data rate over 100 metres of cable:

Category	Data rate
CAT-3	10 Mbit/s
CAT-4	20 Mbit/s
CAT-5	100 Mbit/s
CAT-5e/6	350 Mbit/s
CAT-7	1 Gbit/s

Note 1: 10base-T and 100base-TX transmit over two pairs (one transmit and one receive), thus 10base-T requires CAT-3 cable and 100base-TX requires CAT-5 cable. However, 100base-T4 transmits and receives over four pairs, allowing CAT-3 cable to be used. 1000base-T [1Gbit/s] can be carried over CAT-5e cable by transmitting 250 Mbit/s over each pair.

Note 2: Distances greater than 100 m can be achieved by operating at a lower data rate: i.e. CAT-3 cable can be used to transmit 1 Mbit/s over 250 m.

All categories of cable comprise four twisted copper pairs; each pair being two insulated wires of 0.5 mm diameter (24 a.w.g.) solid wire. The wire insulation affects the transmission performance of the cable and typically PVC is used in CAT-3 and CAT-4 cable, but more expensive Polyolefin is used in CAT-6 or CAT-7 cable. In all cases, the impedance of the pair is about 100 ohms.

The standard colour code for the wire insulation material is as follows:

Pair number	Primary colour	Secondary colour	Jack wiring (TIA 568B)
1	blue	white	4,5
2	orange	white	2,1
3	green	white	6,3
4	brown	white	8,7

The copper pairs are enclosed within a plastic sheath, which is typically made from PVC, Polyolefin or low-smoke/fume (LSF) material. Foiled twisted pair (FTP) cables have an overall metal foil layer inside the plastic sheath; a copper drain wire is provided for ease of connection to an earthing point at the cable termination. Screened twisted pairs (STP) have a metal braid inside the plastic sheath. Unshielded twisted pair (UTP) cables are more common because they are easier to handle and terminate.

9.4 Electrical connectors

A wide range of connector types has evolved for telecommunications applications, including some that have been designed by, and are unique to, British Telecom, but which now appear mainly on older-type equipment. Modern connectors are perhaps most easily considered as falling into four broad categories:

1 Round multipin.
2 Rectangular multipin, including quasi-rectangular types such as D series.
3 Jacks.
4 Coaxial connectors.

In addition to these families of connectors one must also consider the various means by which the connector pins are terminated on their conductor wires. These are: soldering; crimping; wire-wrap; and insulation displacement (IDC). Connectors mainly used in consumer products and power connectors are not included in this book.

9.4.1 Round multipin connectors

Widely used where a rugged, waterproof coupling is required (e.g. in military communications) a range is available to BS 9522 F0017 known as UK Pattern 105 which is compatible with connectors to USA MIL-C-26482. These are bayonet couplings with metal shells and gold-plated copper alloy contacts giving a guaranteed 500 connect/disconnect cycles. Typical ratings are 5 A per contact at 700 V DC or peak AC over a temperature range of −55°C to +125°C. Both chassis-mounting and cable-mounting types are made with numerous pin configurations (e.g. 3, 6, 7, 8, 10, 12, 19, 26 and 32 way).

Lower-cost commercial-quality round multipin connectors are made with a bayonet- or screw-locking arrangement; one commonly available range conforms to MIL-C-5015, has silver-plated contacts and is not guaranteed waterproof. It is typically marketed in 3, 6, 10, 19, 24 and 37 pin versions.

There is also a range of low-cost circular connectors in which the moulded shells are designed to accept pin or socket inserts to which the conductor is connected by a crimped joint (see Section 9.4.5). Pin configurations range from 4 to 48 way, and certain versions comply with VDEO110.

The application of all these multipin connectors is essentially restricted to low-frequency signals.

9.4.2 Rectangular multipin

The three most widely used ranges in telecommunication interfaces are probably the two part moulded indirect edge connectors to DIN 41612/41617, the subminiature 'D' range and the low-cost moulded 'interconnect' series using separate pin or socket inserts (e.g. Molex KK series).

The DIN 41612 series is widely used in 19 inch Euro-card subrack systems for connections between module p.c. boards and the back plane containing the intermodule wiring. Available with wire-wrap or solder spills, and in three classes of performance level with a two-row body giving up to 64 way and a three-row body up to 96 way, these polyester moulded connectors are typically rated at 1.5 to 2 A per pin and a test voltage of 1 kV. The pin spacing is standard 0.1 inch pitch and both straight and 90° terminations are obtainable, as also are mixed contact types incorporating high-power (10 A) contacts and RF coaxial plugs/sockets of 50 ohm impedance, suitable for RG178B/U cable.

The subminiature 'D' range uses polyester moulded contact housings within a trapezoidal metal shell whose shape ensures the correct polarization of male and female connectors. This range is commonly used in RS232/C applications and for generally interfacing computer peripheral equipment. Available in 9, 15, 25, 37 and 50 way format, contacts are typically rated at 5 A, 750 V maximum.

According to price paid, 'D' connectors may be of commercial or industrial/professional quality, the best being to international specifications such as MIL-C-24308 with contact resistance as low as 2.5 milliohms and usable from $-55°$ to $+125°C$.

The range is designed for semi-permanent connections for chassis/cable, cable/cable or cable/p.c.b. interfaces and has a limited guaranteed minimum number of connect/disconnect cycles – usually 50 to 250. 'D' connectors can be obtained with solder, crimp or IDC connections and with straight or 90° connection spills. Their use is restricted to relatively low frequencies, although the availability of IDC (Insulation Displacement Connector) types permits the use of 'twist and flat' ribbon cable which is suitable for some data transmission systems over relatively short distances. 'D' connectors are also available with integral capacitive elements to provide specified attenuation to cable-borne EMI/RFI interference. Typical performance is for 0 to 2 dB attenuation at 3 MHz rising to 20 dB from 100 to 1000 MHz. This conforms to MIL-STD220.

The so-called 'Scart' rectangular connectors appertain to the European audio/video interconnection system between VDUs and computers which provides for audio, video and data cores within one cable/connector system. BS 6552 (1984) specifies the system; the plug and sockets are 21 way and use crimp contacts.

All the above multipin connectors involve the individual termination of each cable wire to a pin or socket, although this may be done semi-automatically with certain crimp connections. A further class of rectangular connectors allows all pin connections to be made simultaneously using the technique of insulation displacement; such connectors are sometimes referred to as 'mass-termination' types. Section 9.4.5 refers to the IDC technique.

IDC connectors are normally used with ribbon cable and are available from several manufacturers as cable-mounting sockets, p.c.b. mounting plugs, edge connectors, dual-in-line (d.i.l.) connectors, transition connectors which allow direct soldered connection of a ribbon cable to a p.c.b., and also 'D' range compatible plugs and sockets. Pin configurations cover up to at least 50 way.

9.4.3 Jack connectors

The jack plug and socket, at one time a familiar part of a manual telephone exchange, is a convenient means of multipole access to a

panel outlet which is still found on some telecommunications installations for auxiliary low-frequency outputs, for example recorders,
loudspeakers.

The cylindrical plug consists of a shaped, brass tip, behind which
may be a number of concentric plated brass rings, insulated from
the tip and each other, and finally a concentric plated brass sleeve
insulated from tip and rings but electrically connected to the body
of the jack plug. The most common is probably the three-way jack
which has just tip, ring and a sleeve which can be connected by solder
joints to the two conductors and screen of a screened cable pair.

The jack socket consists of an equivalent number of sprung contact
pairs spaced within a panel-mounting moulding so as to make contact
with tip, rings and sleeve of the mating plug. The contact pairs may
be normally open or normally closed, thus providing considerable
flexibility of connection between the apparatus and cable.

The standard telecommunications jack plug is 0.250 inches in
diameter and is available with two different profiles to the tip
– known as gauge A and gauge B – which mate only with the
corresponding gauge socket.

Miniature 3.5 mm jacks are used on some consumer equipment.

9.4.4 Coaxial connections

The screened coaxial cables used for radio-frequency transmission
lines are terminated in coaxial connectors which are designed, as far
as is practical, to maintain the characteristic impedance of the cable
at the interface and so avoid unwanted reflections with consequent
loss of power or, in the case of data pulse streams, distortion of pulse
shape or even generation of spurious pulses. All coaxial connections
are engineered to meet the basic requirement of separately terminating
the centre conductor and concentric outer screen. The materials used
depend on the quality and cost of the connector whose size depends
primarily on the RF power to be carried. Most coaxial connector
types have mating male and female versions, but in one category used
for precision laboratory measurements (e.g. of VSWR) the connector
is hermaphrodite and the plane of contact, which is normal to the
cylindrical axis of the connector, is thereby precisely defined.

The termination of RF cable to connector may be in one of
three ways:

1 *Soldering and clamping*: The joint between centre contact and centre conductor is soldered and the flexible outer conductor braid is

clamped in electrical contact with the connector shell by a series of tapered cone washers and nuts.

2 *Crimping*: Both the centre conductor and outer braid are secured by crimping joints. These may be made separately or together but in any case require the use of a special tool. It has been claimed that the VSWR of a crimped connector is lower than that of a similar connector soldered and clamped.

3 Some versions use a combination of soldering and crimping and have a soldered centre conductor and crimped outer.

The most commonly encountered coaxial connections are:

3 mm diameter	SMA 0–12.5 GHz	VSWR $1.20 + 0.035f^*$
		(f = frequency in gigahertz)
	SMB 0–4 GHz	VSWR 1.46 max
	SMC 0–10 GHz	VSWR 1.7 max
	Miniature BNC 0–4 GHz	
5 mm diameter	BNC 0–4 GHz	VSWR 1.2 max
	TNC 0–11 GHz (screw-coupled version of BNC)	
7 mm diameter	N 0–11 GHz	
	C 0–11 GHz	
10 mm (nominal)	Twin axial 0–25 MHz specifically for data pairs	

Coaxial connectors, particularly the smaller types, are available from several sources as cable-mounting (free) plugs and sockets, panel-mounting plugs and sockets, T adaptors, and p.c.b. mounting connectors. Interseries adaptors are also made which permit BNC, N, C and TNC types to be coupled together.

BNC connectors are available in both 50 and 75 ohm versions which are not interchangeable; the other types listed are all 50 ohm.

9.4.5 Connector terminations

Soldering: The traditional method of jointing, it requires the most skill and is labour intensive. The wire may become embrittled behind the joint and stray solder spikes can cause short circuits to adjacent pins.

Crimping: Small flaps on the terminal wire receptacle are forced into a pressure contact with the previously stripped wire, and other flaps are pressed down on the wire insulation to grip it and so prevent the wire pulling free under tension. For consistent results the

---

* Figures for braided cables: the SMA used with semi-rigid (solid outer) cable gives a VSWR of $1.07 + 0.008f$ to 18 GHz.

so-called 'controlled-cycle' hand tool is recommended over simple crimping pliers, even though the tool cost may be 10 times higher. It is essential that the wire is stripped to the dimensions prescribed by the manufacturer and that the proper tool appropriate to the wire size is used. Up to 100 joints/hour can be done manually; up to 4000/hour by semi-automatic machine. Quality control sampling is necessary to ensure consistently reliable joints.

Wire-wrap: The stripped wire is wrapped for two or three turns around the solid square section of the connector pin using a special tool which imparts a relatively constant tension to the wire. The high-pressure contacts at the corners of the square section ensure good conductivity. Both hand and powered tools are available. One advantage of wire-wrap, particularly for prototype work, is that joints can be unwrapped with equal ease using the appropriate tool.

Insulation displacement: The basic principle here is to exert pressure between the two parts of a connector and a ribbon cable, thus forcing the individual connection tines all to pierce the outer insulation simultaneously and make electrical contact with the inner conductors. Although, with care, IDCs can be assembled on to a ribbon cable with no tooling other than a small bench vice, the use of a proper press tool and accessories, available from several IDC manufacturers, is strongly recommended.

Standard ribbon cable has its conductors spaced at 1.27 mm (0.05 inches), is available from 10 way to 64 way and typically rated at 1 A 300 V DC per conductor with a nominal impedance of 92 ohms and a capacitance between adjacent conductors of 56 pF/m. Ribbon cables consisting of alternate lengths (e.g. 450 mm) of twisted pairs separated by a 50 mm length of flat ribbon give improved performance for high-speed data applications, and screened flat ribbon cable, in which an overall wrap of aluminized tape is provided, may be useful where electrostatic coupling to adjacent cables or subassemblies is a problem.

10 Optical fibre cables

Light has been used to transmit messages long before the telephone was ever considered. During Queen Elizabeth I of England's reign (in 1588), the shore line had an array of fires to indicate sighting of the Spanish Armada. A better example is perhaps that of the American Indians who used smoke signals to transmit messages – in digital form! The smoke was amplitude modulated (AM), by placing a blanket over the smoking fire, to produce puffs of smoke. It is, therefore, quite fitting that light modulation should become the solution to high speed, long distance, communications.

The energy carried by light waves having wavelengths between 800 nm and 1600 nm can be transmitted by optical fibre which behaves like a dielectric waveguide. There are three low-loss bands in the spectrum, centred on the wavelengths of 850 nm, 1300 nm and 1550 nm. These bands are known as windows.

Many fibres can be fitted into an optical fibre cable, which is light and easy to handle, and existing cable ducts can be used for installation. Each fibre can carry digital data over a long unrepeated cable length, and at rates of over 1 Gbit/s in the laboratory. By 1997, revenue earning fibres were carrying data at rates of 565 Mbit/s.

10.1 Fibre optic cables

Glass has a refractive index η of about 1.5 and the critical angle θ_c at a glass–air boundary is about $42°(\sin^{-1} 1/\eta)$, above which internal reflection takes place, with equal reflected and incident angles (*Figure 10.1*).

If a second similar boundary, parallel to the first one, is introduced, and the angle of light incidence exceeds θ_{CRIT} light will be propagated along the inner medium (glass) by a series of total internal reflections. Any light wave incident at an angle $\theta_i > \theta$ will propagate along the inner medium; any light incident at an angle $\theta_i < \theta_{CRIT}$ will refract into the outer medium and be lost by scattering and absorption.

If a number of light waves enter the system incident upon the boundary at different angles all greater than θ, propagation will occur over a number of different paths. Because the path lengths differ, TRANSIT TIME DISPERSION occurs and this limits the bit rate that could be transmitted. This type of propagation is MULTIMODE

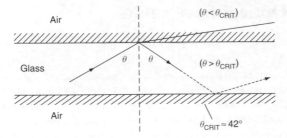

Figure 10.1

PROPAGATION in a STEPPED INDEX fibre, and its use is limited
in practice to low-speed data.

Multimode transmission is minimized by using a GRADED
INDEX fibre in which the refractive index of the inner medium is
made to have a parabolic distribution across a section normal to the
direction of propagation, with a maximum value on the central axis.
Light waves approaching the boundary of the two media will now
be gradually refracted, rather than reflected back towards the centre
axis. Moreover, each change of direction reduces the angle to the
horizontal, and so differences in path lengths are minimized as each
refraction reduces the deviation of the light wave from the centre axis.

An alternative way of eliminating dispersion is to reduce the
dimension of the inner medium (glass core) until its diameter is
the same order of magnitude as the light wavelength so that only
a single mode of propagation is possible. This is STEPPED INDEX
MONOMODE fibre which is capable of very high bandwidth transmission but difficult to manufacture and use in practice.

Practical optical fibres consist of a glass core surrounded by a glass
cladding, both of which must be of high purity. The refractive indices
are of course different, although not necessarily by a large amount – a
10% difference is typical. The next section describes types of optical
cables in common use.

10.1.1 Monomode fibre

The core diameter is of the same order of magnitude as the wavelength of the light to be transmitted, so that only one mode can
propagate.

The core fibre (*Figure 10.2*) will be typically 8 μm in diameter,
but the cladding diameter is standardized at 125 μm. The difficulties
in coupling such a small core to an optoelectronic source/detector or

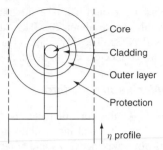

Figure 10.2

to another length of similar fibre are an obvious disadvantage, but precision fibre splicing tools and connectors have produced low-loss joints. This type of fibre is now common in public telephone networks, linking trunk switching centres to main telephone exchanges.

Light of different wavelengths travels at different speeds and this effect is called chromatic dispersion. Chromatic dispersion causes light pulses to be spread over time and limits the data rate, since the pulses merge together, and is due to the finite spectral width of an optical source. Chromatic dispersion is a minimum at 1300 nm, but the minimum can be shifted to coincide with the 1550 nm window. Dispersion shifted monomode fibre is produced by modifying the centre core of the fibre, so that the refractive index of the core has a W profile, rather than a simple step.

10.1.2 Stepped index multimode

The core diameter is typically 50 μm (*Figure 10.3*) but can be larger and the cladding diameter is 125–150 μm (Ref. ITU-T G651); their refractive indices may differ by only 10%.

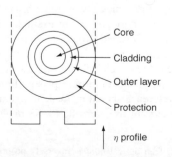

Figure 10.3

110

Such fibres are used for short-haul (i.e. 1–2 km) low-cost installations with bandwidths up to about 20 MHz. Applications include local area networks (LANs). The advantage of this type of fibre is that cheap, easy to fit, connectors can be used.

10.1.3 Graded index multimode fibre

Here again (*Figure 10.4*) the core diameter is 50–60 μm and the cladding diameter 125 μm. Graded index fibres are in some parts of the world used both for trunk networks and some local networks at 0.85 and 1.3 μm wavelengths as they are relatively cheap to produce and capable of bandwidths up to 1 GHz/km compared with 20 MHz/km* for stepped index fibres. The core has its maximum refractive index at the centre decreasing parabolically towards the core/cladding boundary.

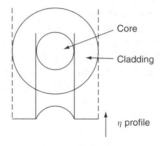

Figure 10.4

10.1.4 Optical fibre cables

Optical fibre cables, carrying a number of fibres, are usually built up around a steel wire core which gives strength, and may also contain copper conductors to supply power, for example to series-fed digital regenerators. Fibres are inserted into plastic tubes for protection before being enclosed in an aluminium foil, or similar water barrier, wrapped around to form the cable which is finally sheathed overall with an age-resistant jacket of polythene or similar plastic. Individual fibres may be colour coded; reference should be made to manufacturers' data for details. Cables in current use carry 8, 48, 88 or 96 fibres, in 8-fibre bundles.

* The bandwidth of an optical fibre is inversely proportional to its length; figures of GHz/km or MHz/km are sometimes called 'bandwidth efficiency'.

10.1.5 *Optical attenuation*

It is essential to realize that an electrical measurement of 2 dB is equivalent to an optical measurement of 1 dB since optical power is proportional to current, but electric power is proportional to (current)^2. Therefore

$$\text{dBm (optical)} = 10 \log \frac{P_{\text{OUT}}}{1 \text{ mW}}$$

and

$$\text{dB (optical)} = 10 \log \frac{P_{\text{OUT}}}{P_{\text{IN}}} = 10 \log \frac{I_{\text{OUT}}}{I_{\text{IN}}}$$

Light energy fed into a fibre suffers attenuation due to absorption, scattering in the core by inhomogeneities in η (Rayleigh scattering), scattering at the core boundary, imperfect input/output and joint couplings, and radiation loss at bends in the fibre.

Typical attenuation of monomode fibre is 1 dB/km at a wavelength of 850 nm and 0.3 dB/km at 1300 nm. The lowest loss occurs at 1550 nm, where typical attenuation is 0.2 dB/km.

10.2 Couplers, connectors and splices

Two main types of demountable couplers are in use for multimode fibre: end-fire and lens couplers.

End-fire coupling is the situation where both optical faces are parallel and in close proximity. Insertion loss is a minimum when both faces are the same diameter (symmetrical end-fire coupling). It is large when the exit face is larger than the entrance face (e.g. LED to fibre) but low when the exit face has the smaller diameter (e.g. fibre to photodiode).

Lens coupling uses optical convex lenses to couple an exit face to an entrance face, and although this allows greater mechanical tolerances at couplings, optical attenuation is higher than with end-fire coupling which is therefore the preferred method.

Attenuation of a coupling is affected by:

(a) axial misalignment of fibres
(b) angular misalignment of fibres
(c) reflections at glass/air interface (Fresnel loss)
(d) end-face roughness.

Monomode fibre connectors are commonly SMA or FC/PC types. A variant on the FC/PC is the FC/APC which has an angled face. The 15° angle prevents reflections from the face being transmitted back towards the source. This type of connector is used with linearly modulated lasers, to reduce noise caused by reflections.

The majority of optical fibre connections are made by fusion splicing. This involves removing the nylon outer coating from the fibre ends, exposing the bare glass. The fibre ends must then be cleaved to give a right-angled surface on each fibre. An adhesive filled plastic tube is slid over the end of one fibre for use after splicing is complete. The fibres are then inserted into a fusion splicing tool where the fibre ends are butted together with careful alignment and an electric arc is struck across the join to melt the glass. The two fibre ends are pushed slowly together, while surface tension helps to align the splice. The adhesive fitted plastic tube can now be slid over the splice and heated so that the tube shrinks as the adhesive melts, sealing the glass in a protective cover. Bare glass is made brittle by water molecules in the air, so this last stage is vital.

10.3 Optical transmitters

Two types of source are in use to provide light energy at the appropriate wavelength and which can be modulated at high data bit rate.

The LED is used for short-haul narrow-bandwidth systems and the laser diode for long-haul wideband applications. The latter provides up to 100 times the output of an LED but is more expensive and less reliable.

The LED produces non-coherent light energy and is suitable for graded index multimode fibres. Choice of semiconductor material determines the wavelength of transmitted light. Power output is typically 0.05–1 mW at wavelengths of 0.8 to 1.35 μm (800–1350 nm).

The laser diode is essentially an LED incorporated into an optical cavity which is resonant at the transmitted wavelength. At low current the behaviour is like a normal LED, but above a threshold value simulated photon emission occurs due to optical resonance, and the light output rises rapidly. The light energy is coherent, that is the light waves are in phase, travel in the same direction, and are virtually all the same wavelength. Power output is typically 5–10 mW and the spectral width 3.5 nm.

Laser diodes are normally always used with monomode fibres, and can be modulated to gigahertz bandwidth. They are also extremely sensitive to drive current above the threshold value and often require cooling and power stabilization control.

10.3.1 External modulators

Amplitude modulation of a laser can be achieved by switching it on and off, but the rate at which this can be done is limited by the time needed by the laser to reach the maximum light output power. Also, turning a laser on and off produces light that is not spectrally pure.

Lasers produce high levels of spectrally pure light when driven by a constant d.c. signal. An external modulator, such as an electro-absorption modulator (EAM), varies its optical attenuation in proportion to the electrical signal being applied across it. This allows the data rate to be 2.5 Gbit/s, or more, whilst maintaining a spectrally pure signal. This is important if wavelength division multiplexing is to be used (see Section 10.5).

10.3.2 Optical fibre safety

The human eye is susceptible to damage from laser light and therefore care must be taken when handling the optical fibre used by a working system. NEVER look at the end of a fibre connector of a working system using a microscope or magnifying glass. Indirect viewing should be used, using a camera fitted to a microscope.

Optical safety levels were revised during 2000. Most light sources are Safety Class 1 or Class 1M (formerly Class 3A). Class 1 has a 15 mW power limit at 1300 nm and a 10 mW limit at 1550 nm. All LED devices and some lasers come into this category; they are low-power and inherently safe (Class 1). Class 1M covers lasers up to 50 mW power limit at 1300 nm and approximately 150 mW at 1550 nm. These power levels are regarded as safe for 'live working' but only when precautions are taken, as described above.

Handling coated optical fibre is normally quite safe, but handling bare fibre is hazardous because glass fibre becomes brittle when exposed to air. It is quite easy for fibre to pass through the skin, resulting in uncomfortable glass splinters that are very difficult to remove. The handling of bare fibre is normally only necessary when carrying out fusion splicing and this should be done carefully for both safety reasons and for splice quality reasons.

10.4 Optical detectors

A semiconductor photodiode is used to convert photons to electrical energy, and consists of a reverse-biased pn junction in which the current flowing is almost directly proportional to the illumination on the junction and virtually independent of the bias voltage so long as the junction remains reversed. In the absence of illumination the small reverse current flowing (dark current) is a limiting parameter in determining receiver signal/noise performance.

Two types of diode are in use – the PIN photodiode and the avalanche photodiode. The PIN diode has an intrinsic (i.e. neutral) layer sandwiched between the p and n layers and this effectively increases the switching (modulation) speed by reducing the junction capacitance. PIN diodes may be silicon, germanium or indium/gallium/arsenide and have both good linearity and high bandwidth.

The silicon avalanche photodiode operates in a reverse break-down mode and therefore internally amplifies the photocurrent. The gain/bandwidth is sufficient to convert optical power in the range of milliwatts at frequencies of up to 1 GHz, but avalanche diodes are less stable than a PIN diode followed by an electronic amplifier.

Optical fibre amplifiers are described in Chapter 18.3.

10.5 Wavelength division multiplexing (WDM)

The data carrying capacity of an optical fibre is immense, but is lim-ited in most transmission systems by the electronics used to drive the laser. As described in Section 10.3.1, using a modulator exter-nal to the laser allows rates of 2.5 Gbit/s. Higher data rates can be transmitted over a single fibre by transmitting several light sig-nals simultaneously. It is necessary for each light signal to have a different wavelength (i.e. a different colour), so that they can be separated at the receive end. This is wavelength division multiplex-ing (WDM).

In a WDM system there are a number of light sources feeding into a fibre at one end, and a number of photo-detectors at the other end. A typical number of transmitters and receivers is 16. Such a system requires good separation between light signals at both ends of the fibre. At the transmit end, a directional coupler is needed so that light output from one laser does not enter into another and disrupt its lasing action. At the receive end, photo-detectors are sensitive to a wide range of wavelengths and a coupler incorporating wavelength filtering is required to ensure that light of unwanted wavelengths does not reach the detector.

The ITU-T has produced a list of wavelengths that give wave-length separations of between 100 GHz and 1 THz, as shown in

Table 10.1 ITU-T recommended wavelengths for WDM systems

Wavelength (nm)	100 GHz	200 GHz	300 GHz	400 GHz	500 GHz	600 GHz	800 GHz	900 GHz	1 THz
1530.33	195.9	X							
1531.12	195.8		X						
1531.9	195.7	X		X			X		
1532.68	195.6								
1533.47	195.5	X	X		X	X			X
1534.25	195.4								
1535.04	195.3	X		X					
1535.82	195.2		X					X	
1536.61	195.1	X							
1537.4	195				X				
1538.19	194.9	X	X	X		X	X		
1538.98	194.8								
1539.77	194.7	X							
1540.56	194.6		X						
1541.35	194.5	X		X	X				X
1542.14	194.4								
1542.94	194.3	X	X			X		X	
1543.73	194.2								
1544.53	194.1	X		X			X		
1545.32	194		X		X				
1546.12	193.9	X							
1546.92	193.8								
1547.72	193.7	X	X	X		X			
1548.51	193.6								
1549.32	193.5	X			X				X
1550.12	193.4		X					X	
1550.92	193.3	X		X			X		
1551.72	193.2								
1552.52	193.1	X	X			X			
1553.33	193				X				
1554.13	192.9	X		X					
1554.94	192.8		X						
1555.75	192.7	X							
1556.55	192.6								
1557.36	192.5	X	X	X	X	X	X	X	X
1558.17	192.4								
1558.98	192.3	X							
1559.79	192.2		X						
1560.61	192.1	X		X					
1561.42	192				X				
1562.23	191.9	X	X			X			
1563.05	191.8								
1563.86	191.7	X		X			X		
1564.68	191.6		X					X	
1565.5	191.5	X			X				X

Table 10.1. The first column gives the wavelength in nanometres, the second column gives the frequency in THz for channel spacings of 100 GHz. Subsequent columns indicate frequencies that can be used to give channel spacings of 200 GHz, 300 GHz, 400 GHz, etc., up to 1 THz spacing.

11 Radio

11.1 Radio propagation

The advent of electro-magnetic interference (EMI) regulations has made people realize that it is not difficult to transmit radio signals through the atmosphere, indeed it is difficult to stop them. The Earth's surface is conducting, and tends to reflect radio signals. So too are layers of the atmosphere, called the ionosphere, where sunlight has caused molecules to break down and become ionized (conducting). Needless to say, the day-time performance of the ionosphere is different from night-time, and sunspot activity can change the ionosphere too.

The radio spectrum is divided into bands. Each band has different propagation characteristics:

- The very low frequency (VLF) band is used to transmit signals at slow speeds to submarines, since only VLF signals travel a few metres through the sea's surface.
- The low frequency (LF) band covering 300 kHz to 3 MHz is used for local radio because the range is short.
- The high frequency (HF) band covers 3 MHz to 30 MHz; signals in this range have global reach, since the radio waves bounce between the ionosphere and the Earth's surface.
- Very high frequency (VHF), ultra-high frequency (UHF) and microwave bands cover 30–300 MHz, 300 MHz–1 GHz and >1 GHz respectively. These signals have 'line-of-sight' propagation. They do, however, reflect off buildings and other objects which improves their coverage of the area around the transmitter. Signals requiring a higher bandwidth, such as broadcast television, are generally assigned higher frequency bands.

11.1.1 Antennae

A radio frequency signal injected into an unterminated piece of wire will act as an antenna (sometimes called an aerial) and allow electro-magnetic radiation. If the piece of wire is exactly a quarter wavelength long, and perpendicular to the Earth's surface, the radiation is more efficient. In simple terms, current from the signal source travels along the wire at the speed of light, but is reflected from the open circuit at the wire's end. The reflected signal coincides with the signal source changing polarity and the voltages of the transmitted and reflected waves add. The wire thus forms a resonant circuit, producing a large electric field along its length. Radiation efficiency is increased by the resonance of the antenna. This type of antenna is called a monopole. The Earth's surface can be replaced by any large conducting surface, such as a vehicle or aircraft body.

A balanced signal can be radiated efficiently from a dipole, which is two monopoles pointing in opposite directions. The field strength of the radiation is greater than for a monopole and has the advantage of allowing the field plane to be changed from vertical to horizontal.

Antenna can be made directional, thus giving a gain when compared with the dipole. This is achieved by adding wire elements in front of, and behind, the dipole at approximately quarter wavelength spacing. The elements in front of the dipole are known as directors; these are slightly shorter than a half wavelength, and the number of elements determines the gain. There is usually a single element, called a reflector, which is longer than a half wavelength and is fitted behind the dipole. This antenna is called a Yagi, after the Japanese engineer who invented it.

Monopole, dipole and Yagi antennae are illustrated in *Figure 11.1*.

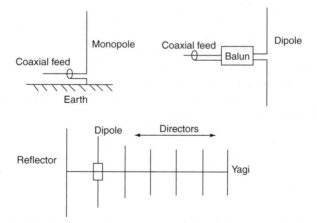

Figure 11.1 Wire antennae

Dish antenna are used at microwave frequencies. The parabolic reflector makes the antenna highly directional, so alignment between transmit and receive antennae is important.

11.2 Local access

Radio links are sometimes used for local access, instead of using copper pairs. This could be where the terrain is difficult, such as in mountainous areas, or where a second telephone services provider wished to compete with an existing company. Microwave frequencies (above 2 GHz) are usually used for these services, partly because of the total bandwidth requirements and partly because the lower frequencies are already in use. Digital modulation methods (such as time division multiple access) are used which allow new services to be provided as they are developed, and hence give the second telephone service provider a marketing 'edge'.

11.2.1 Wireless local loop (WLL)

The local loop is the copper connection between the local telephone exchange (central office) and the customer. Unbundling of the local loop, to give competing companies access to the copper pair, has taken place in several parts of the world. In some circumstances, access to an existing copper pair is not practical and installing new cable is too expensive. In these circumstances a Wireless Local Loop (WLL) may be viable. There is not a 'standard' WLL, instead a number of possible systems are employed. The most popular narrow-band systems are DECT, PHS and CDMA (frequency hopping or direct sequence), which have been discussed in Chapter 4. Broadband local access systems (MMDS and LMDS) are discussed later in this chapter.

The DECT system was developed in Europe and operates in the 1880 to 1900 MHz frequency band. Signals are transmitted in short bursts at a power level of 250 mW. Channel separation is by using both FDMA and TDMA techniques: the carrier frequency and time slot are dynamically allocated when in standby mode or during calls. Each system has a choice of 10 carriers and 12 time slots. The send and receive channels are separated by time (time division duplex, or TDD). DECT uses ADPCM speech coding to give a 32 kbps data stream, which is then encrypted before transmission. Handshaking takes place between the base unit and the remote unit to give additional security. The system operates over a line-of-sight range of 6.5 km, but is generally limited to below 6 km due to TDMA timing conditions.

The personal handy-phone system (PHS) was developed in Japan and is very similar to DECT. It uses the same 1880 to 1900 MHz frequency band and uses ADPCM speech coding to give a 32 kbps data stream, which is then encrypted before transmission. The advantage of PHS over DECT is the operating range, which is 13 km. A variation on PHS is Personal Access Communications System (PACS).

In the US and Russia, CDMA is a popular choice for WLL systems. Details on how CDMA works were given in Chapter 4. A number of CDMA-WLL systems are available and they use voice coding schemes that have data rates from 13 kbps to 32 kbps. As previously mentioned, CDMA requires accurate power control of all the transmitters within a cell. For mobile systems this was a challenge, but in fixed systems the radio path loss does not change very much and so can be handled relatively easily.

11.3 Cellular networks

Cellphones are radio transceivers used by people who are away from their office or home telephone. Quite often these people are moving, either on foot or in a vehicle. Consequently, the cellular network is designed to allow for movement by having a number of base stations linked into the fixed telephone network; as the cellphone user moves from one base station's area to another, the radio link is switched between them. This involves re-tuning the cellphone to a different channel, as well as switching the fixed network from one base station to the other.

Cellular networks using analogue modulation or digital time division multiple access (TDMA) modulation occupy the UHF and microwave bands (there are several bands above 800 MHz and below 2 GHz). Each base station employs multiple transmitters to allow for several calls in progress at one time. Adjacent cells use different frequencies, but cells further away can re-use frequencies since the 'line-of-sight' range is short.

An exception to the multi-frequency cellular network is where the digital code division multiple access (CDMA) modulation is used. This modulation technique allows many channels to operate over the same carrier frequency. In CDMA, each user is assigned unique codes to replace the digital 1s and 0s that represent the speech. The bandwidth is greatly increased since a string of 1s and 0s must be sent for each single bit of the original, however all the cellphones can operate at the same frequency since each person's code is different. The advantage to the operator is that only one radio transmitter (and receiver) is required at each base station.

11.4 Microwave links

Fixed microwave links are used by telecommunication service providers and the military for trunk links between different areas, often to give diverse routing of important circuits. Towers with several microwave antenna mounted on all sides can be seen when driving in many countries (see *Figure 11.2*). These operate in several frequency bands, within the 3 to 30 GHz range.

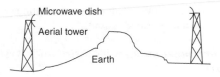

Figure 11.2 Microwave link

Television companies use microwave links for outside reporting. A van with a microwave antenna mounted on the roof is used in conjunction with a television camera to transmit images from the site of incidents, such as major accidents or natural disasters like earthquakes. Quite often the television picture is transmitted via a satellite link, since line of site is not usually possible.

11.4.1 Local multipoint distribution system (LMDS)

LMDS provides broadband communications in the 'local-loop'. A central site broadcasts on a one-to-many basis, whilst the remote sites can signal the central site on a one-to-one basis. Data rates of 36 Mbps on the shared down-link are available, with up-link rates from a few kHz up to 8 Mbps is typical (25.8 Mbps maximum). Thus LMDS is ideal for interactive digital television.

Physically, a system consists of two primary functional layers: transport and services. The transport layer comprises the customer premises roof-top-unit (RTU) and the node electronics. The RTU solid-state transceiver is approximately 30 cm in diameter. The node electronics includes solid-state transmitters, receivers and other related elements. The services layer comprises a network interface unit (NIU) at the customer premises and the base electronics. The NIU provides ATM or PDH-based access interfaces to the customer. The base electronics provide control and transport functions from the central hub-site.

The frequency spectrum allocated by the Federal Communications Commission (FCC) for local multi-point distribution services (LMDS) totals 1350 MHz. The allocation is in four separate blocks of spectrum with two located in the 28 GHz band (27.5–28.35 GHz and 29.1–29.25 GHz), one located in the 31 GHz band (31.075–31.225 GHz), and one in the 40 GHz band (40.5–42.5 GHz). The 40 GHz band could be extended to 43.5 GHz. In these frequency bands, the attenuation of water (rain, condensation, fog, snow and tree foliage) has a significant effect on the transmission range and transmission is over line-of-sight paths only. The line-of-sight restriction for LMDS has the largest impact on system planning and coverage is limited to a 3–5 km range. Transmission from the antenna is sectored into four quadrants of alternating polarity, which allows effective re-use of the spectrum.

11.4.2 Multi-channel multi-point distribution system (MMDS)

The multi-channel multi-point distribution system (MMDS) provides users with data rates of up to 10 Mbps. These data rates are much lower than those available using LMDS, but MMDS operates in the 2.5–2.7 GHz frequency band and has an operating range of up to 45 km. By using lower radio frequencies, MMDS suffers less from the effects of rain and humidity.

The narrow bandwidth allocation (200 MHz) is forcing manufacturers to investigate methods of reusing the frequency band. One method is to sector the antenna, so that transmission is made directional. The same frequency can then be used to transmit in another direction without causing significant interference.

11.5 Satellite communications

11.5.1 Orbits

A body is in a stable circular orbit around the earth when the radial acceleration due to gravitational pull is equal to the radial acceleration resulting from the curved trajectory. When this condition is satisfied the force due to gravity on the body is balanced by the centrifugal force, resulting in the phenomenon of weightlessness.

If the angular velocity of the body is ω radians per second, and the radius of the orbit (i.e. the distance from the centre of the earth) is r kilometres, then

$$\text{Acceleration due to gravity} = \frac{39.5 \times 10^7}{r^2} \text{m/s}^2$$

$$\text{Radial acceleration due to circular path} = \omega^2 r \times 10^3 \text{m/s}^2$$

Hence for stable orbit $\dfrac{39.5 \times 10^7}{r^2} = \omega r \times 10^3$

$$\omega = \frac{\sqrt{39.5} \times 10^2}{r^{3/2}}$$

$$\text{Orbit time} = 2\pi \frac{r^{2/3} \times 10^{-2}}{\sqrt{39.5}} = 10^{-2} r^{3/2}\,\text{s}$$

The mean radius of the earth is 6378 km; hence the height of the trajectory above the earth is

$$h = r - 6378 \text{ km}$$

The orbit time in hours, plotted against h, is shown in *Figure 11.3*. Orbits may occur in any direction across the earth's surface provided the plane of the orbit passes through the earth's centre. Low-level satellites have a short orbit time of the order of 1.5 hours, and with a polar orbit, the earth will have rotated by 22.5° during one orbit. Such orbits enable the earth's surface to be scanned in successive orbits, 16 being required to cover the whole surface. Careful adjustment of the orbit time enables different strips to be surveyed on successive days; for instance, for the example quoted, in 22 days the whole of the earth's surface could be surveyed at 1° intervals. Such orbits are useful for aerial surveillance purposes, but of very limited use for telecommunications owing to their short availability time and the need to track their movement.

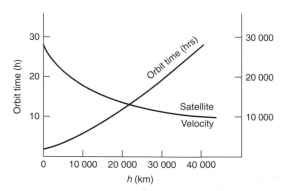

Figure 11.3 Circular orbit time

Figure 11.3, however, shows that a critical value of h exists for which the orbit time is 24 hours. If a satellite is put into an equatorial orbit at this height it will circulate the earth at the same angular

124

velocity as the earth's rotation, and when viewed from the earth it will appear to be stationary. Such satellites are known as geostationary or geosynchronous satellites. Their advantages for telecommunications – 24 hour availability with no requirement for tracking – are obvious.

In one year the earth makes 366 rotations in free space and one orbit of the sun, resulting in a sidereal day of 23 hours, 56 minutes and 04 seconds, which is the orbit time required for a geostationary satellite. The value of h for this orbit time is 35 788 km.

The disadvantage of the high value of h is the transmission delay, which for a minimum return path of 72 000 km is 240 ms. The advantages can be seen from *Figure 11.4* showing the earth and the orbit to scale.

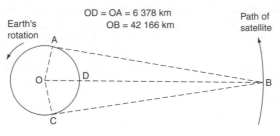

Figure 11.4 Geostationary satellite

From this diagram:

$$A\hat{O}C = 162.6°$$

$$A\hat{B}C = 17.4°$$

$A\hat{O}C$ shows the section of the earth's surface where the satellite appears to be on the horizon, the theoretical limit of range. On the same line of longitude as the satellite this represents a coverage from latitude 81.3° N to latitude 81.3° S, and along the equator, 81.3° of longitude east and west of the overhead position. The diagram also shows that the high value of h results in a reduction of the time in which the satellite is in the earth's shadow, rendering the solar cells aboard inoperative. From *Figure 11.4* it would appear that this would occur for 17.4° of the orbit, representing shadow for a period of 70 minutes in each day. Owing to the inclination of the earth's axis to the plane of rotation around the sun (23°) this state only applies when the equatorial plane is in line with the sun, that is at the spring and autumn equinoxes. On either side of the equinoxes the period of shadow progressively decreases, and beyond 23 days the inclined axis takes the orbit outside the earth's shadow and the solar cells remain in sunlight for the full 24 hours of each orbit. Some satellites are located to the west of their service area to bring the blackout period to the small hours of the morning when the power demands are at their lowest.

Once in orbit, the satellite is clear of the earth's atmosphere and friction losses are very small. There are, however, second-order effects that influence the orbit path, the most significant being the irregularities in the shape of the earth, which is not a perfect sphere, and the gravitational pull of the moon and sun. The position of a geostationary satellite therefore requires careful monitoring, and periodic corrections have to be applied by remote signals from the earth controlling station.

The velocity of the satellite in circular orbit is equal to ωr. Hence:

$$\text{Velocity} = \frac{\sqrt{39.5 \times 10^2}}{r^{3/2}} \times r \times 3600 = \frac{2.26 \times 10^6}{\sqrt{r}} \text{ kph}$$

Velocity is plotted against height in *Figure 11.3*. To launch the satellite it is first carried to a height of about 300 km by means of a rocket or space shuttle. Orbit speed at this height is approximately 27 000 kph, and the satellite has to acquire a higher velocity to escape the low orbit. If the velocity is increased to 36 700 kph, the satellite will go into an elliptical orbit with a height at apogee of 36 000 km, but with a velocity at apogee reduced to 4800 kph. If launched from a rocket the additional speed required to change from circular to elliptical orbit can be obtained by increasing the rocket velocity before release. If launched from a shuttle, release takes place at orbiting speed, and the additional speed has to be obtained from a perigee boost motor in the satellite, reducing the payload of the satellite. For this reason, most launches into the high geostationary orbit utilize rockets, the shuttle being used for low-orbit military and surveillance satellites.

When the apogee of the elliptical orbit is at the correct height for a geostationary orbit, an apogee thruster in the satellite is fired to accelerate to the correct orbit velocity of 11 200 kph, at which speed the orbit will become circular.

11.5.2 Satellite consortia

The costs of manufacturing and launching a satellite are very high.

Economic considerations therefore limit the providers of satellite services to government-sponsored or privately owned corporations, or consortia. Consortia have been formed to satisfy a number of different markets that satellites are capable of serving:

(a) *Mobile*: Satellites provide communication over wide areas to mobile terminals. This needs to be an international operation and the international agency INMARSAT has been formed to provide and operate the satellites. Initially INMARSAT have confined their activities to shipping.

(b) *Intercontinental*: Satellites provide communication over a wide area to fixed terminals. This enables permanent speech and data links to be set up between countries, forming part of the international public service network. Again this requires an international consortium, leading to the formation of INTELSAT. Approximately 120 countries are members of INTELSAT, with around 180 countries or territories making use of its services.

(c) *Limited area*: Land masses such as Europe, Australia, North America, the Near East, etc. can justify the use of limited-area satellites to provide links between countries or territories within the area. This has led to the formation of a number of consortia to satisfy local demands independent of INTELSAT. The principal consortia are:

Europe	EUTELSAT
USA	WESTAR
	RCA SATCOM
	SATELLITE BUSINESS SYSTEMS
	GTE SATELLITE CORPORATION
Canada	TELESAT
Middle East	ARABSAT
India	INSAT
Australia	AUSSAT
Indonesia	PERUMTEL

(d) *Television*: Geostationary satellites have opened up the opportunities for wide-area television. Transmission may be either through a dedicated satellite owned by the television company, or by hiring part of the output capacity of a consortia-owned satellite. Television, like radio broadcasting, is part of the entertainment industry and outside the scope of this book.

(e) *Private*: As an extension to (c), a large or multinational company may find that its volume of interbranch communications justifies the ownership of its own satellite, particularly if much of the communication is long distance.

11.5.3 Ancillary satellite equipment

The useful payload of a satellite is its communications equipment, but additional equipment has to be fitted to position and operate it. An apogee boost motor and fuel is required to change from an elliptical to a circular orbit (Section 11.5.1), and if launch is from a space shuttle a perigee boost motor is also required.

When in orbit the satellite must remain directionally stable, with the side bearing the aerials facing the earth, and any rotation controlled to keep the sides facing in the required direction. Without correction this would not be achieved, and deviations have to be detected by sensors and corrected either automatically, or by instruction from the controlling earth station. Corrections are applied by means of small thrusters, so located that any attitude correction can be obtained by selection of the appropriate thrusters. The actual position of the satellite relative to earth will also be subject to drift. Each satellite is allocated an assigned place, which it should maintain within $\pm 0.1°$ as seen from the earth, an accuracy sufficient to avoid the need for any realignment of the fixed earth-station dishes. The same correction thrusters can be utilized to alter the satellite's position when required. The satellite therefore has to be equipped with a number of thrust motors, and sufficient fuel for its design lifespan (of the order of 10 years). The signalling of information relating to the state of the satellite, and instructions to the satellite, require a telemetry link between the earth control station and the satellite.

The other vital piece of ancillary equipment is the power supply. Power is obtained from solar cells, usually mounted on wings outside the main body of the satellite. These wings may need to be folded during launching and opened when in orbit. They also need to be rotated once in 24 hours in order that the cells will always face the sun. The sun's radiation provides 1400 watts of energy per square metre, but the conversion efficiency of currently available solar cells is only of the order of 10%, resulting in a generation of electrical power of about 140 watts per square metre. Available satellites have a power capacity in the range 1–7 kilowatts. Loss of power for up to 70 minutes in 24 hours around the spring and autumn equinox (Section 11.5.1) creates the need for battery backup. Only some 20–30% of the power is radiated as transmitted signal, the remainder being dissipated as heat in the satellite. This heat, together with the heat collected on the surface facing the sun, has to be lost by heat radiation, and further equipment may be necessary to transfer heat to the cold side of the satellite.

11.5.4 Satellite communications equipment

A satellite receives weak signals from earth, and retransmits at much higher power back to earth. The close proximity of receive and transmit aerials would lead to instability if the satellite acted as a repeater or regenerator. Retransmission therefore has to take place at a different frequency, and frequency changers are interposed between receivers and transmitters, forming a series of transponders. The 'up' frequency is normally higher than the 'down' frequency, and the frequency changer is often referred to as a 'down convertor'.

Each transponder consists of a filter and low-noise pre-amplifier or receiver followed by the down convertor, high-power transmit amplifier and an output filter coupling into the aerial. The low-noise pre-amplifier can be of conventional solid-state design with a flat response over a frequency range of the order of 500 MHz.

It is, however, essential to design the transmit equipment to obtain adequate signal strength at the required earth terminals. This involves the optimum division of the available power between the transmitters, which may have to cover widely different areas, the provision of high-efficiency transmitters and the careful design of the aerials to concentrate the signal on the required sites. Output power required can vary between 3 and 400 watts, the lower powers being used for narrow beams into specific small areas, and the higher powers for wide continental coverage. At the lower frequencies solid-state amplifiers are employed, but at the higher frequencies travelling wave tubes (TWTs) have to be used. These are relatively inefficient, converting only about 30% of the input power into radiated signal. More than one amplifier may be required to produce the output power.

Directional microwave transmission is obtained by using parabolic dish reflectors. The characteristics can be expressed in terms of beam angle and gain. Beam angle is defined as the solid angle in which the majority of the beam power is focused, and is taken to the −3 dB boundary of the beam; hence the smaller the beam angle the smaller the area covered at the earth's surface, but the higher the signal strength in this area. Gain is defined as the power per unit area in the beam relative to the power produced by spherical radiation. The surface area of a spherical surface d kilometres from the radiator is $4\pi d^2$, and a transmitter radiating W watts will produce $W/4\pi d^2$ μW per square metre on this surface.

This figure will be increased by the gain of the reflector; hence the gain of the reflector is a vital factor in minimizing the required transmitted power. The values of beam angle and gain are given by

$$\text{Beam angle (3 dB boundary)} = 21/fD \text{ degrees}$$

$$\text{Reflector gain} = 60(fD)^2 \text{times}^*$$

$$= 10\log_{10} 60(fD)^2 \text{dB}$$

where f is the frequency in gigahertz and D is the dish diameter in metres.

The value of beam angle and gain are plotted against fD in *Figure 11.5*. A value of fD of 10 would give a beam angle of 2.1° and a gain of 37.8 dB, and would apply to a frequency of 5 GHz and diameter of 2 metres, or a frequency of 10 GHz and diameter of 1 metre.

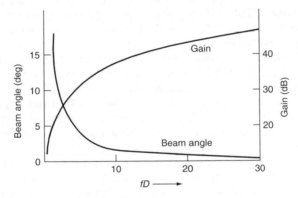

Figure 11.5 Parabolic reflector characteristics

Doubling either frequency or diameter would halve the beam angle.

The signal strength in microwatts per square metre at the earth's surface can therefore be estimated from the transmitted power and beam angle or aerial gain. For comparative purposes the signal strength produced by a satellite transmitter is expressed in terms of the 'equivalent isotropic radiated power' (EIRP), which is the sum of the aerial gain in decibels and the radiated power in decibels relative to 1 watt, that is an aerial system with a gain of 40 dB and radiated power of 10 W would have an EIRP of 50 dBW. Attenuation due to the earth's atmosphere and heavy rain become significant above 10 GHz, and have to be taken into account.

* Assuming 55% efficiency.

A single receive aerial will frequently feed several transponders, and the output from the amplifiers of several transponders will feed a single transmit aerial. Transponder bandwidths can vary over a range of 30–500 MHz. Using frequency-modulated speech channels a single transponder handling a 70 MHz band would carry approximately 1000 channels. A satellite having five receive and transmit aerials, each pair of aerials handling the traffic of five transponders, would therefore have a capacity of the order of 25 000 speech channels. A simplified block diagram is shown in *Figure 11.6* for three aerial systems each handling three transponders.

Aerial system 1 utilizes up signals in the 14 GHz band and down signals in the 11 GHz band; aerials 2 and 3 up signals in the 6 GHz band and down signals in the 4 GHz band. All incoming signals are down-converted to the 4 GHz band before passing through the transponders, and for the high-band signals a further up conversion is required to the 11 GHz band. Since all signals are converted to the 4 GHz band before passing through the filters and amplifiers in the transponders, outputs can be cross connected, for example a transponder connected to receive aerial 1 could be connected to the output amplifier of a transponder feeding aerial 3. If, therefore, each of the receive aerials is focussed on different areas, say Western Europe, the Middle East and North Africa, and each transmitting aerial is beamed to the same area, cross connection of the transponders will provide communication between each area and the other two.

To increase flexibility in the utilization of the links, replacement of the fixed links by remotely controlled switching is now common practice. Crosspoint switching has the advantage that it can switch both analogue and digital signals. Also, by using double-changeover coaxial switches the switching can be carried out on the 4 GHz transponder signal. This is shown in *Figure 11.6* with the three 3 × 3 crosspoint switches.

The arrangement shown in *Figure 11.6* implies the use of FDM for the radio signals, each aerial handling the traffic for several transponders in separate frequency bands, which are separated in the input band-pass filters. FDM has the advantage of simplicity, which is of particular importance if some of the land stations are in the hands of private users or unskilled personnel. It enables undigitized speech to be transmitted using FM, and data using simple FSK or some of the more complicated FSK and PSK derivatives, depending on customer facilities and requirements. The amount of frequency band allocated to a particular channel can be varied to suit requirements (provided it does not exceed the transponder bandwidth), and it can cope with a

131

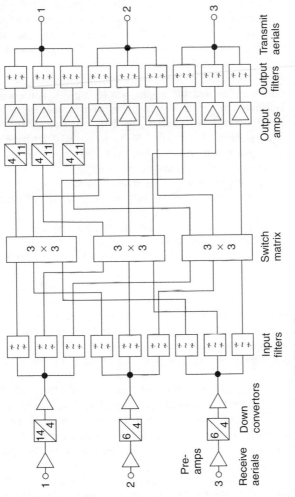

Figure 11.6 Satellite transponders – simplified block diagram

wideband analogue television transmission. A disadvantage of FDM is that the output amplifiers have to be operated below maximum power to reduce intermodulation products.

The increasing amount of data transmission, the replacement of FDM trunk links by TDM, and the ultimate change of the entire telecommunication system to ISDN is leading to the wider adoption of digital techniques in satellites. Rather than modulate the carrier with a single high bit rate TDM signal, TDMA (Time Division Multiple Access) can be applied, whereby the signals to be combined, and subsequently separated into their individual transponders, are each allocated a time slot, and utilize the whole bandwidth for a short period. Transmission can then take place from a number of separate locations provided they synchronize their time slots. This system requires compression of the signal to be transmitted into the time slot duration, which can be carried out relatively easily with a digital signal, but not with its analogue counterpart. The system also requires more complex terminal equipment, including the provision of accurate synchronization between transmitter, satellite and receive station.

Digital transmission does, however, provide the possibility of digital switching of time slots, both time switching to change the slot position and space switching to switch to different output amplifiers.

Both systems of switching dealt with above switch the whole band of a transponder, which may contain 1000 channels. To switch smaller groups of channels, or individual channels, involves demultiplexing bit streams to lower orders of TDM, or baseband signals. This provides a further increase in the flexibility of the satellite, and with digital signals and LSI, is now a worthwhile option for many applications.

With the exception of some data transmissions, and television relay, satellite communication is two-channel duplex. The very high-level differences between transmit and receive aerials result in a degree of cross-talk. The frequency difference between up and down transmissions enables most of this to be suppressed by the input filters. Residual interference will be returned with a time delay to the station where the transmission originated, and will therefore appear as an echo. Many satellites are now equipped with 'echo cancellors', whereby a cancelling signal is inserted in the return path to suppress the echo.

The frequency bands used in Region 1 (Europe) are summarized in *Table 11.1*.

Table 11.1

Frequency range (GHz)	Up/down	European telecom. links	International telecom. links	Other uses
1.5–1.6	Down	Mobile		
1.6–1.7	Up	Mobile		
2.5–2.6	Down			Broadcast*
3.4–4.2	Down	Fixed	Fixed	
4.5–4.8	Down	Fixed	Fixed	
5.9–7.0	Up	Fixed	Fixed	
7.2–7.7	Down			Military
7.9–8.4	Up			Military
10.7–11.7	Down	Fixed		
11.7–12.5	Down	Fixed		Broadcast*
12.5–12.75	Down	{ Fixed		
12.75–13.25	Up	{ private links		
14.0–14.8	Up	Fixed		
17.3–18.3	Up	Fixed		
17.7–20.2	Down	Fixed		
20.2–21.2	Down	Mobile		
22.5–23.0	Down			Broadcast*
27.0–30.0	Up	Fixed		
30.0–31.0	Up	Mobile		
22.55–23.55				
32.0–33.0				
54.25–58.2		Allocated for intersatellite links		
59–64.0				

* Broadcast includes television links

11.5.5 Satellite land stations

Mobile land terminals require small and simple aerial systems, and their satellite transmitters have to cover wide areas and produce strong signals within the service area, requiring high-power transmitters.

As pointed out in Section 11.5.4 satellite links can be one way or two way, and earth stations may be required for either one- or two-way operation. The directional and gain properties of dishes are similar for both transmit and receive. With a fixed earth station pointing at a geostationary satellite, narrow-beam high-gain reflectors can be used which after alignment can remain fixed.

An earth station operated by a PSO and carrying a large number of international channels can justify the use of sophisticated equipment. This can include large dishes and low-temperature preamplifiers to reduce thermal noise. The increase in size and power of modern satellites enables these installations to be simplified, and a single large station replaced by a number of smaller stations distributed in high call density areas in a country.

Smaller earth terminals are required for private links, which may be installed on the roof of an office block belonging to a multinational company, and provide a 'private line' facility via the satellite to other company sites in other countries or continents.

The simplest earth terminal is that required for receive only at domestic premises for television reception. This requires a small inconspicuous dish which can be easily aligned on the satellite and continue to operate without further attention. The simplification of the earth terminal, combined with the need for a wide coverage area for the transmission, has to be paid for in terms of transmitted power, which may be in the order of 400 watts.

The complexity of the earth receive equipment required will decrease as the EIRP increases. A wide-area transmission producing an EIRP of less than 35 dBW will require a sophisticated receiving system, probably operated by a PSO. EIRPs in the region of 45–50 dBW can be received on simpler public or company-owned receiving systems, but an EIRP exceeding 55 dBW is required for the small dish aerials used for satellite television.

11.5.6 Intersatellite links

Geostationary satellites can communicate with other geostationary satellites or with low-orbit satellites without transmission through the atmosphere.

As shown in Section 11.5.1, the maximum span on the equatorial plane for a satellite is from about 80° E to 80° W of the satellite. To obtain transmission beyond this distance involves a hop to another satellite either via earth, or by direct satellite-to-satellite transmission. The latter alternative has the advantage of bypassing the earth station.

A low-orbit satellite moves position relative to the earth and can only maintain communication contact for a short period during each orbit unless a series of earth stations around its orbit are available. It remains visible to a geostationary satellite for a much longer proportion of its orbit, and information can be transmitted from a low-orbit satellite to a geostationary satellite and then back to a fixed earth station. With surveillance satellites this enables much more data to be transmitted to the earth station.

Frequencies allocated for intersatellite communication are included in *Table 11.1*.

11.5.7 Satellites for mobile communications

INMARSAT are operating a series of satellites for maritime communications. This includes four INMARSAT 2 and five INMARSAT 3 satellites covering the Atlantic, Pacific and Indian Ocean areas. The system has been developed to provide basic speech and low-speed data with relatively simple equipment on the ships. The main data requirements are telex and the ability to receive weather maps, both of which can be satisfied by low bit rates. Facilities are also provided for distress calls and navigation.

It is expected that satellite facilities will be extended to mobile vehicle communications, providing international vehicle-to-vehicle and vehicle-to-base links, and also possibly to paging. The third sphere of mobile satellite communications is expected to be aeronautical, augmenting the radio facilities already available to aircraft.

Frequencies allocated for mobile communications with satellites are given in *Table 11.2*.

Table 11.2

		Frequency (MHz)
Maritime	Ship to satellite	1626.5–1645.5
	Satellite to ship	1530–1544
	Satellite to shore	4192.5–4200
	Shore to satellite	6417.5–6425
Land mobile	Satellite to earth	1530–1533; 1555–1559
	Earth to satellite	1631.5–1634.5;
		1656.5–1660.5
Aeronautical	Satellite to earth or aircraft	1545–1555
	Earth to aircraft to satellite	1646.5–1656.5

The mobile services listed in *Table 11.2* require a terminal the size of a laptop computer, to which a moderate sized dish antenna is attached – a far cry from the cellphone.

The demand for hand-held mobile phones that can be used anywhere in the world has given rise to the development of low earth orbit (LEO), or medium earth orbit (MEO), satellite systems. The satellites in these orbits are not geostationary; they have to circle the Earth much faster than the rotation rate of the Earth itself, in order to stop themselves falling to earth. Therefore, LEO and MEO satellite systems require a much higher number of satellites, 66 in the case of the LEO Iridium system, because each one is only over a country for a short time. A summary of the systems deployed or planned is given in *Table 11.3*.

11.5.8 Satellites for ground-station communications

A geostationary satellite can be seen from the earth across an arc of 163° (Section 11.5.1), at the ends of which the satellite appears on the

Table 11.3 Mobile satellite systems

System	Orbit	Number of satellites
Aries	LEO	48
AMSC (USA only)	MEO	10
Ellipso	LEO/MEO	16
Globalstar	LEO	48
Iridium	LEO	66
ICO Global	MEO	10
Odyssey	MEO	12

horizon. Allowing for a minimum elevation of about 6°, this reduces to approximately 150°. *Figure 11.7* shows these limits for three strategically placed satellites A, B and C, which together cover most of the earth's continents and oceans in three regions, Atlantic, Indian and Pacific. To cover the entire service area with one receive and one transmit aerial would require a beam angle of 17° (*Figure 11.4*), providing a gain of approximately 19 dB (Section 11.5.4). Very high transmitter power would be required in the satellite, or an elaborate earth receiving station, to obtain a satisfactory signal, and a large proportion of the transmitted power would be wasted on sparsely populated areas and oceans. Normal practice is therefore to utilize several aerial systems directed towards areas of dense population. A conical beam, with a beam angle of θ degrees will diverge to a circle of approximately $630 \times \theta$ kilometres diameter at the earth's surface. This will produce a circular pattern immediately below the satellite, but as the elevation angle decreases, the pattern becomes elliptical, and the area covered increases. A 1° beam, focused on to a heavy-traffic area, will have a gain in the order of 44 dB, requiring only 1/300 of the power required for the 17° beam, and careful targeting of the aerials is required in order to minimize the power requirements.

International links are operated by INTELSAT using over 20 high-powered satellites. INTELSAT V, V(a), VI, VII and VII(a) series satellites are all in operation, giving worldwide coverage. A single satellite, INTELSAT K, offers Ku band coverage of the Atlantic Ocean area. By the end of 1998, the INTELSAT VIII series of six satellites will be operational.

Satellites are positioned to provide capacity where there is a demand. For example the INTELSAT 6 series has five satellites, three are positioned over the Atlantic Ocean and two are above the Indian Ocean. The coverage of these satellites is described in *Figure 11.7*. For satellite A, one set of aerials is directed to America, and a second

137

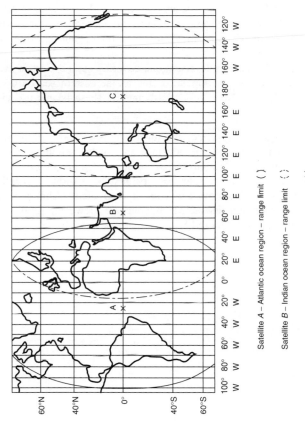

Satellite *A* – Atlantic ocean region – range limit ( )

Satellite *B* – Indian ocean region – range limit ⟨ ⟩

Satellite *C* – Pacific ocean region – range limit ⟨ ⟩

Figure 11.7 Maximum coverage for geostationary satellites

set to Europe, the Mediterranean and Africa, with reuse of frequencies between the two areas. Three types of beam are employed:

1 *Hemi-beam*: A wide beam, one covering the whole of Europe, the Middle East and Africa, and the second the eastern side of North America and the whole of South America, both operating in the 6/4 GHz bands.
2 *Zone beam*: A smaller-area beam, one covering Europe, the Middle East and North Africa, and the second the eastern side of North America and the Central American states, both again operating in the 6/4 GHz bands.
3 *Spot beam*: A small-angle beam, one covering central Europe and the second the densely populated area on the eastern seaboard of the USA (Boston, New York, Washington). These beams use the 14/11 GHz bands.

This arrangement provides for links between nearly all of the major cities within the overall coverage area of the satellite, with a capacity relating to the anticipated traffic density on each link.

A variety of different satellites operated by various consortia (Section 11.2) provide coverage over limited areas. The general principles of operation and aerial targeting are, however, similar to those employed for the INTELSAT international satellites.

12 Radio modulation

Radio frequencies can be transmitted over long distances, but carry no information unless they are modulated. There are several ways of modulating a carrier signal, and the mathematical analysis of these techniques is beyond the scope of this book. The modulating signal (i.e. the data we want to transmit) is known as the baseband.

1 *Amplitude modulation (AM)*, where the modulating signal controls the signal level of the carrier. This can be achieved by multiplying the modulating signal voltage and the carrier signal voltage together; this is known as mixing. In the frequency domain, mixing translates the baseband spectrum and its mirror image to either side of the carrier frequency, as shown in *Figure 12.1*. The frequency shifted baseband signal and its mirror image are known as sidebands (since they are either side of the carrier). The upper sideband occupies frequencies above the carrier, whilst the lower sideband occupies frequencies below the carrier.

2 *Frequency modulation (FM)*, where the frequency of the carrier signal increases and decreases and the voltage of the baseband signal rises and falls. The resultant spectrum is more complicated than for AM and occupies far more bandwidth. This modulation technique is non-linear, so harmonics of the baseband signal are produced and translated to either side of the carrier frequency. The amount of deviation of the carrier frequency relative to its centre frequency is a measure of the modulation depth. The greater the modulation depth, the greater the number of baseband harmonics generated and, consequently, the greater the bandwidth required. Broadcast radio stations usually occupy a bandwidth of ± 75 kHz, for a baseband bandwidth of 15 kHz.

3 *Phase modulation (PM)*, where the phase of the carrier is changed. This method is related to frequency modulation and, for a sinewave, the result is identical. The difference between them can be seen if a digital modulating signal is applied, because the phase of the carrier switches to its new value and produces a step change in carrier amplitude. A variation on PM is phase shift keying (QPSK) where in-phase and quadrature carrier signals are switched by the data. In some cases the data stream is filtered so that the phase shift is not instantaneous.

4 *Quadrature phase and amplitude modulation (QAM)*, where both phase and amplitude are changed. This modulation method is used

140

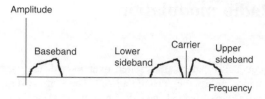

Figure 12.1 Spectrum after mixing

with digital systems and each phase/amplitude value can represent several data bits. The number of states is given in the description, thus 16QAM has 16 states, each representing 4 bits of data. In the case of digital video broadcast (DVB), the replacement for analogue broadcast television, 64QAM will be used to reduce the bandwidth requirements (each state represents 6 bits).

Other modulation schemes are based on those described above. Spread spectrum, for example, modifies the baseband signal before it modulates a carrier using QPSK. In direct sequence spread spectrum, each data bit is converted into a predetermined sequence of 10 or more bits. The bandwidth requirement is higher, but several channels can operate at the same frequency and the risk of interference is lower.

The radio spectrum is crowded, so accuracy of the carrier frequency is important. This accuracy is also important for the interoperability of telecommunications systems. The methods of achieving accuracy are discussed in Section 12.1.

12.1 Signal sources

Modern telecommunications place some stringent requirements on the frequency resolution and stability of signal sources: for example, the FDM system requires carrier frequencies and pilot tones of high stability in both amplitude and frequency; the PCM system requires high-stability clock frequencies; and narrow-band filters and the like require high-resolution, stable signal sources for test and maintenance. Radio communication links also require close control of carrier frequency to prevent interchannel interference.

Three types of frequency standard are in common use:

1 Caesium atomic-beam-controlled oscillator.
2 Rubidium gas-cell-controlled oscillator.
3 Quartz crystal oscillator.

12.1.1 Atomic standards

The caesium standard depends on a naturally invariant frequency and is an international primary standard. The second is defined as 9 192 631 770 periods of transition within the caesium atom. In practice a caesium frequency standard uses this as a master clock from which is synthesized a range of standard frequencies such as 10 MHz, 5 MHz, 1 MHz, etc. The quoted accuracy of the Hewlett Packard 5061B/004 caesium standard is $\pm 7 \times 10^{-12}$ from 0 to 50°C with a long-term stability of $\pm 2 \times 10^{-12}$.

The rubidium standard, like caesium, uses an atomic resonance to control a quartz oscillator via a frequency lock loop, but is subject to a small drift since its stability depends on gas mixture and pressure in the rubidium gas cell. The rubidium cell is a secondary standard subject to periodic calibration and typically provides a long-term drift rate better than 1×10^{-11} per month.

12.1.2 Quartz standards

The quartz crystal standard is inexpensive compared with atomic standards, and with proportional control of crystal oven temperature can achieve short-term (1 s) stability of 5×10^{-12}, RMS long-term ageing of 5×10^{-10} per 24 hours, and a temperature stability of about 2×10^{-9} over the range of 0–50°C.

12.1.3 Synthesizers

Where the stability associated with the above is not required, simpler forms of *LC* and *RC* oscillator may be used, but the facility of the 'fractional N' type of frequency synthesis to generate a whole range of frequencies from one reference source has led to its increasing use. This system allows a single phase-locked loop to generate at least 10 digits of frequency resolution for frequencies integrally and non-integrally related to the reference frequency. The elementary phase-locked loop forming the basis of a synthesizer is shown in *Figure 12.2*.

The output from the phase detector (see Section 12.2), which is proportional to the phase difference between its inputs, is filtered in a low-pass filter and used to bias a varactor diode which controls the frequency of the oscillator, which in this case is that of the reference input. The latter is crystal controlled and so the output will be of the same stability apart from any noise introduced within the loop. By interposing programmable frequency dividers between the voltage-controlled oscillators and phase detector the output can be set

142

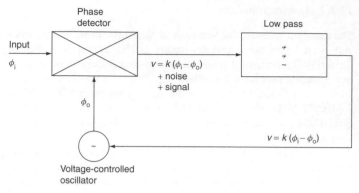

Figure 12.2

to a large number of different frequencies all having a stability comparable with that of the reference frequency. In particular a technique called dual modulus prescaling, in which the first divider following the VCO can be programmed to divide by N or $N + 1$ according to the contents of one of a second pair of dividers, allows a wide range of division ratios in steps of one. This change of one is made to represent the required frequency increment (e.g. a system channel separation of 5 kHz) by the designer's choice of VCO frequency and division ratios. Commercial examples of test sources based on this principle, and including AM/FM modulation facilities, cover, in total, from DC to 26 GHz. The long-term stability is that of the crystal reference frequency, but synthesis produces some phase 'noise' at frequencies very close to the carrier.

12.1.4 LC and RC oscillators

By contrast, the free-running LC and RC oscillators which synthesizers have largely replaced produce stabilities no better than that of the inductors, capacitors and resistors which determine the operating frequency. In practice this is in the order of 0.1% to 1% per 24 hours at nominally constant temperature. The undoubted advantages of these oscillators are their comparative simplicity and ability to produce low-distortion sinusoidal outputs without recourse to additional filtering.

12.1.5 Off-air frequency standards

The principal industrial countries in the world have at least one broadcast radio station transmitting a frequency which is controlled by an atomic clock. The carrier may also be modulated with timing information, for example seconds, minutes, hours, etc.

In the UK the national standard is MSF Rugby on a frequency of 60 kHz with an effective radiated power of 27 kW; in the USA, WWVB broadcasts on the same frequency. Both stations broadcast continuously apart from a regular maintenance 'outage' for a few hours each month (MSF Rugby 10.00 to 14.00 hours on the first Tuesday).

MSF Rugby is controlled by a caesium beam frequency standard with an accuracy of ± 2 parts in 10^{12}; WWVB has an uncertainty of 1 in 10^{11}. The time signals from Rugby are carrier breaks of 100 ms for the second pulses and 500 ms breaks for minute pulses, the start of the pulse indicating the epoch. A binary code of 10 ms bits inserted in the minute slot (fast code) gives time of day, month and day of month. A second binary code at 1 bit/s (slow code) from seconds 17 to 59 gives time of day, day of month, day of week, month and year information.

Also in the UK, BBC Radio 4 transmissions on 198 kHz from Droitwich, Burghead and Westerglen are controlled to ± 2 parts in 10^{11} by rubidium gas-cell standards. These signals carry phase-modulated time of day and calendar information in addition to amplitude-modulated programme material.

World-wide a number of different frequencies are used for standard frequency transmissions, notably Allouis (France) 162 kHz, Moscow 4.996, 9.996 and 14.996 MHz, Rome 5 MHz, and Mainflingen (Germany) 77.5 kHz.

A number of manufacturers market radio receivers designed specifically to provide timing standards locked to a standard frequency transmission.

12.2 Modulators and demodulators

Examples of circuit techniques for modulators and demodulators are given in this section. There are many variations of these circuits in existence; more often these circuits are included within an integrated circuit or hybrid module, with no access to the individual components.

12.2.1 Amplitude modulation

The basic formulae for amplitude modulation are summarized in Chapter 26. Many circuit configurations have been used to amplitude-modulate a carrier signal. For example, since the current gain (h_{f_e}) of a transistor depends on the collector current, modulation of the latter will produce an AM output in the simple circuit in *Figure 12.3*.

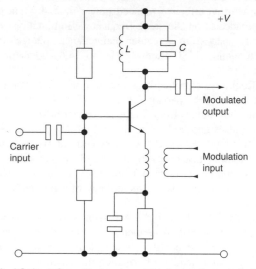

Figure 12.3 *LC* circuit *Q* must be low enough to give band pass including modulation sidebands

The linearity of a modulator will be improved by the use of feedback; if the AM output is detected in a low-distortion detector, its envelope may be compared with the original modulating signal, and so the former is forced to follow the modulating waveform (*Figure 12.4*).

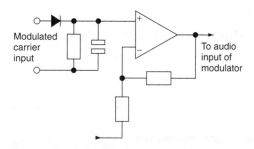

Figure 12.4

A high-powered modulator for broadcast transmitters may well use thermionic valves; in a typical example a class B triode push–pull

stage amplifying the modulating signal would supply, via a transformer, the anode voltage for a carrier amplifier valve feeding the
aerial. The latter would operate in class C with a resonant anode load
and the modulation supply to the anode (equal to $2 \times$ DC supply)
could be in the order of kilovolt peak to peak.

A PIN diode behaves like a resistor at frequencies greatly above
$f_c = 1/2\pi\tau$, where τ is the lifetime of a minority carrier. By varying
the bias the RF resistance can be made to vary typically between 1
and 10 ohms, and this makes the PIN diode useful as a modulator (or
attenuator, phase shifter or AGC element) at frequencies from about
1 MHz to at least 18 GHz. Below f_c the device behaves as a normal
diode.

The examples given above all produce a double-sideband output
plus carrier. Some radio communication systems and all FDM carrier systems require single sideband (SSB) or SSB with suppressed
carrier. The unwanted modulating signal, carrier and one sideband
can be removed by filtering. The carrier itself can, alternatively, be
suppressed by using a balanced form of modulator (e.g. the balanced
transistor modulator shown in *Figure 12.5 (a)*).

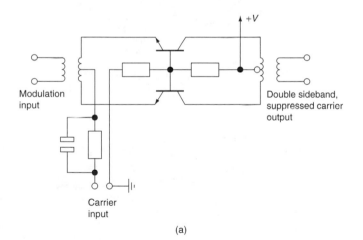

(a)

Figure 12.5(a)

The ring modulator (*Figure 12.5 (b)*) and Cowan modulator
(*Figure 12.5 (c)*) use fast diodes as switches operated by the carrier
signal. The modulation signal, of lower amplitude, flows through
the diodes only during the half cycle of carrier that makes them
conduct.

146

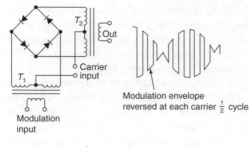

Output = $KE_m \cos\omega_m t \cdot \frac{4}{\pi}[\sin\omega_c t + \frac{1}{3}\sin3\omega_c t + \frac{1}{5}\sin5\omega_c t + ...]$

Wanted OP of $\omega_c \pm \omega_m$ Unwanted OP of $3\omega_c \pm \omega_m$, $5\omega_c \pm \omega_m$...

(b)

Figure 12.5(b) T_1, T_2 are balanced (e.g. bipolar) transformers

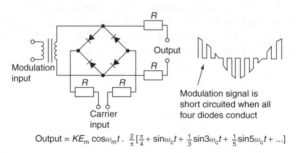

Modulation signal is short circuited when all four diodes conduct

Output = $KE_m \cos\omega_m t \cdot \frac{2}{\pi}[\frac{\pi}{4} + \sin\omega_c t + \frac{1}{3}\sin3\omega_c t + \frac{1}{5}\sin5\omega_c t + ...]$

(c)

Figure 12.5(c)

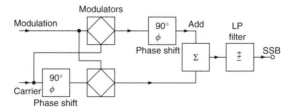

Figure 12.6

Circuits have been devised for cancelling one sideband; British patents 260 067 and 586 278 and US patent 2 248 250 refer to these. The basic principle, shown in *Figure 12.6*, demands the use of a phase-shifting network with a flat amplitude versus frequency characteristic and 90° phase shift over the sideband spectrum.

This circuit is commonly referred to as the Weaver single side-band modulator. A variation on this circuit, using four mixers and two quadrature local oscillator sources, is used to demodulate single side-band signals and is known as the Barber receiver. The mathematical analysis is beyond the scope of this book.

12.2.2 Frequency modulation

The basic formulae for frequency and phase modulation are summarized in Chapter 26. Frequency modulation is achieved by varying one of the timing elements of an oscillator; in an RC timed circuit either the resistance or capacitance (usually the former) is varied in sympathy with the modulation whilst in an LC oscillator it is the inductance or capacitance that is so varied.

For example, FSK modulation used in data communications is easily achieved by switching alternative values of resistors in an RC timed circuit using diodes or FETs to provide the switching function.

It is usual to use a varactor diode as a voltage-controlled capacitor to modulate an LC oscillator at frequencies between about 500 kHz and several hundred megahertz. One example used at 10 MHz is shown in *Figure 12.7* where C_4 is a DC blocking capacitor and C_2 and C_3 are chosen to be much larger than C_1; then frequency is determined by L and C_1 with varactor diodes D_1, D_2.

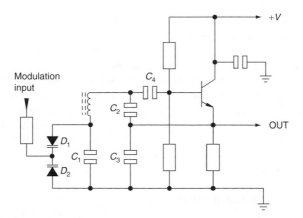

Figure 12.7

Two diodes are used back to back to avoid damping due to forward conduction, but this does reduce frequency deviation because of the series connection. Capacitors C_2, C_3 swamp the intrinsic (and variable) transistor capacitances. The centre (unmodulated) frequency can be stabilized by comparing the mean frequency to a crystal reference in a phase-sensitive detector and using the DC error as part of the bias to the varactor diodes.

Alternatively and particularly where a wide deviation is required (e.g. 5–15 MHz), the inductor L may be made as a current-controlled inductor. That is, a second winding on the same closed ferrite core is biased with a modulating current to control the core permeability and hence the value of L.

The voltage-controlled oscillator, compared with a reference frequency in a feedback loop, is the basis of the frequency synthesizer. This subject is dealt with in Section 12.1.

The frequency deviation of an oscillator is increased by frequency multiplication and, conversely, reduced by frequency division. The percentage deviation can be increased by frequency multiplication followed by frequency changing in a mixer to restore the carrier to its original value, but this process will naturally also proportionally increase unwanted modulation due to noise, 50 Hz hum, etc.

12.2.3 Phase modulation

Small amounts (e.g. $\pm 10°$) of phase modulation can be achieved by varying the capacitor or inductor of a tuned buffer amplifier following the unmodulated carrier frequency oscillator. A better arrangement usable to at least $\pm 30°$ with less than 5% distortion is to sum the output of a balanced amplitude modulator (Section 12.2.1) with a 90° phase-shifted version of the carrier, finally passing the result through a limiter stage to remove amplitude fluctuations.

A phase-modulated signal can also be obtained using any of the previously described methods of frequency modulation but with the modulating voltage or current first passed through a network which makes its amplitude proportional to its frequency, for example a simple CR differentiating network. As with frequency modulation, if a phase-modulated signal is passed through a frequency multiplier or divider, the modulation index is increased or decreased in proportion.

12.2.4 Vector modulation

Microwave and data communications are making increasing use of modulation techniques involving the transmission of digitally encoded levels and phases. Vector modulation involves two separate orthogonal quadrature (Q) components. Some modems use QAM (Quadrature

Amplitude Modulation) whilst digital microwave radio currently uses both QAM, PSK (Phase Shift Keying), and multilevel (*m*-level) modulation. The object is to increase the number of bits per baud above unity, hence increasing the bits per cycle of bandwidth.

In a vector modulator, the carrier is divided into in-phase and quadrature signals and mixed with the modulating *I* and *Q* signals. The mixer outputs are summed to produce the final vector-modulated output (*Figure 12.8*).

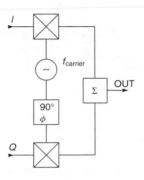

Figure 12.8

12.2.5 AM detectors

The diode detector can be used in two forms as shown in *Figure 12.9*.

In *Figure 12.9 (a)* the diode conducts only on the peaks of the positive-going cycles of carrier frequency, since $CR > 1/f_c$. The modulation envelope is, in effect, sampled, once every cycle of the carrier waveform. The input level must be large enough to switch the diode at the modulation trough, and the value of R should be large compared with the forward resistance of the diode, but small compared with its reverse resistance. C must not have a low impedance at the highest modulation frequency, and the following ratio of impedance should be observed for minimum distortion of the recovered modulation waveform:

$$Z_m/R \gg \text{modulation depth (at highest modulation frequency)}$$

where Z_m is the impedance of C in parallel with R at the modulation frequency. In *Figure 12.9 (b)*, with a shunt diode connection, the capacitor C gives DC isolation between input and output and the demodulated output is clamped to 0 volts (ignoring the diode drop) at its positive peaks. The filter components R_1C_1 are necessary to

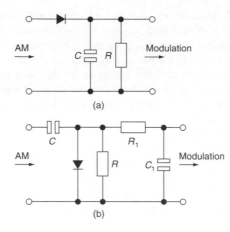

Figure 12.9

remove residual carrier frequency from the output, the attenuation being $1/2\pi f R_1 C_1$, when $C_1 R_1 \gg 1$.

A different form of detector is required for transmission systems in which the carrier frequency and one sideband is suppressed; here a signal, synchronous with the carrier, is developed at the receiver and used to switch diodes in a balanced circuit. Two examples are shown in *Figure 12.10*.

A product detector can also be designed using transistors in an emitter-coupled configuration in which the common collector current is controlled by one input, whilst the second input is applied differentially to the two transistors. Since transistor gain is a function of collector current a multiplicative action occurs so that if one input is in the form $\cos \omega_c t$ and the other $\cos(\omega_c \pm \omega_m)t$ (i.e. carrier and one modulation sideband) the output will contain $\cos \omega_m t$ and higher frequencies which can be filtered off. This circuit is the basis of many integrated circuit multipliers and synchronous detectors.

In addition to detection/demodulation of AM, there is a requirement, particularly in the measurement field, to convert AC signals to an equivalent DC value for driving a meter or digital display. The peak detectors of *Figure 12.9* may be used, or a mean level detector as shown in *Figure 12.11*, or some form of true RMS detector. The most basic RMS detector is the vacuum junction thermocouple in which the unknown voltage energizes a small heater, and a thermocouple detects its temperature rise. Heater and thermocouple are mounted in close proximity in a small evacuated glass envelope. Disadvantages are low-impedance input (typically a few ohms) and comparatively

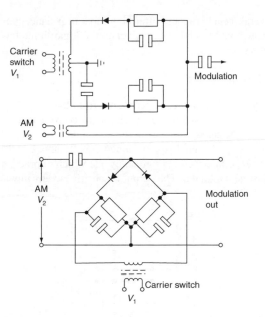

Figure 12.10 $V_1 > V_2$, which is insufficient to cause diodes to conduct; V_1 is synchronous with the AM carrier frequency

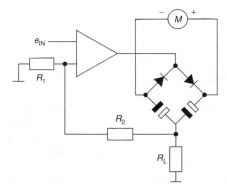

Figure 12.11

long time constant, although these can to some extent be overcome by use of negative feedback amplifiers. The prime advantage, other than true RMS law, is their accuracy over a frequency band up to VHF.

Integrated circuit RMS detectors, using a totally different principle based on an algorithm, are now widely used for measuring instruments up to several megahertz, and offer high input impedance, temperature-compensated output, and even conversion to logarithmic law for decibel displays.

12.2.6 FM detectors

Frequency modulation may be recovered from the carrier by using the characteristic of a resonant tuned circuit to convert it to AM which can then be detected by one of the methods of Section 12.2.5. In *Figure 12.12*, L_1C_1 and L_2C_2 are tuned equally above and below the mean carrier or intermediate frequency f_c, which in practice is not an easy adjustment to make. The output is sensitive to any unwanted AM.

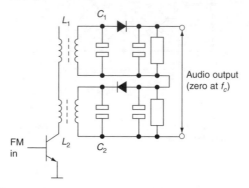

Figure 12.12

An improved form of FM detector, or discriminator, is the circuit ascribed to Foster and Seeley and shown in *Figure 12.13*. Here the transformer secondary, tuned to f_c, is centre tapped and a sample of the primary voltage is injected into it. At resonance the voltage across the secondary is in quadrature with the primary voltage. At frequencies (above/below) resonance the secondary voltage (lags/leads) the primary and V_1, V_2 vary in accordance with the frequency deviation, as shown by the vector diagrams in *Figure 12.14*.

By reversing one diode in the preceding circuit we have a ratio detector in which a degree of AM rejection is obtainable (*Figure 12.15*). The output across $C_2 + C_3$ remains constant, but the voltage across either capacitor varies according to its frequency modulation. By using a relatively low value for R_1 and making C_4R_1 about 1 s, a degree of AM rejection is obtained.

A totally different method of detecting frequency modulation is to use a pulse-counting technique; one such circuit is the diode pump,

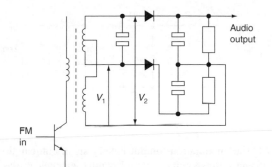

Figure 12.13

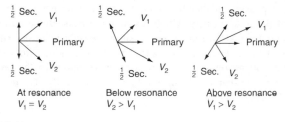

Figure 12.14

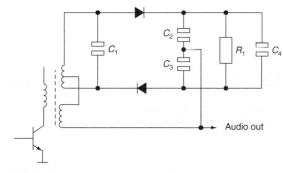

Figure 12.15

which needs an amplitude-limited pulse input. In *Figure 12.16*, each time the input goes negative C_1 charges to V_{IN} through D_1; each time V_{IN} returns to 0 volts, C_2 acquires charge via D_2. If C_2R is large a DC voltage appears across R proportional to the input pulse repetition frequency and is in fact $V_{IN}C_1Rf$ providing $C_1Rf \ll 1$, that is providing $V_{OUT} \ll V_{IN}$.

154

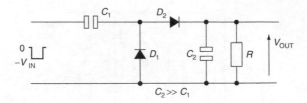

Figure 12.16

The AC (i.e. modulation) output is very small compared with the DC component corresponding to mid band frequency in the normal case where the latter is much greater than the deviation. Hence this type of demodulator requires a low IF prior to demodulation.

An improvement can be made to this circuit which eliminates much of the non-linearity which restricts the basic diode pump to a small output signal. In *Figure 12.17*, since the right-hand side of C_1 is 'caught' at V_{OUT} (less V_{be}) by the transistor, D_1 conducts immediately the input voltage returns to earth after every pulse. In the previous circuit the right-hand side of C_1 had to rise to $V_{IN}(C_1/C_1 + C_2)$ before D_1 could conduct.

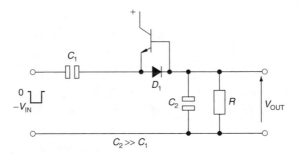

Figure 12.17

A further form of FM detector uses the principle of the phase-locked loop (*Figure 12.18*). The VCO must be linear if the correction voltage applied to the VCO is to represent the undistorted demodulated signal.

At high input signal/noise ratio there is little to choose between a phase-locked loop and conventional discriminator. At low levels of S/N ratio the improvement gained by using a PLL depends on the modulation index, being the greater for high values of m. If possible

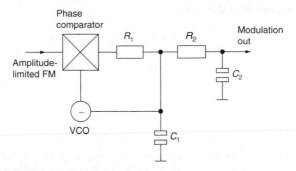

Figure 12.18

the loop and filter should be designed specifically for the anticipated modulation.

12.2.7 Phase detectors

Any ideal analogue multiplier behaves as a phase detector. If Input 1 = $V_1 \sin(\omega t + \theta_1)$ and Input 2 = $V_2 \sin(\omega t + \theta_2)$, the product $V_1 V_2 \sin(\omega t + \theta_1) \sin(\omega t + \theta_2)$ contains the term $\frac{1}{2} V_1 V_2 \sin(\theta_1 - \theta_2)$, and if $\theta_1 - \theta_2$ is small, the output is proportional to $(\theta_1 - \theta_2)$ since $\sin(\theta_1 - \theta_2) \approx (\theta_1 - \theta_2)$.

However, the typical analogue multiplier is normally restricted to use at relatively low frequencies and for high-frequency systems a switching-type phase detector is used. The switch can be a transistor, or a diode bridge (*Figure 12.19*).

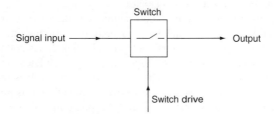

Figure 12.19

If the signal input is $V_1 \cos(\omega t + \theta)$ and the switch drive changes state at the zero crossings of ωt, the output is

$$\begin{aligned} &V_1 \cos(\omega t + \theta) &&\text{for } \theta < \omega t < \pi \\ &0 &&\text{for } \pi < \omega t < 2\pi \end{aligned}$$

The mean level of the output is

$$\frac{V_1}{\pi} \int_0^\pi \cos(\omega t + \theta)\, d(\omega t) = -\frac{V_1}{\pi} \sin\theta$$

The linearity can be much improved if both inputs are square waves, for then the output, instead of being a function of $\sin\theta$, is triangular, passing through zero at $-\pi$, 0 and π. So even for phase angles as large as 90° linearity is maintained, and this type of phase detector is often used in PLL frequency discriminator.

An alternative way of looking at this phase detector, with square-wave inputs, is as an exclusive-OR gate whose output is the time average of its two logic states. As a phase detector this system has the disadvantages of being limited to a phase shift below $\pi/2$ (modulation index <1.5), and requiring a switch drive frequency (unmodulated).

This same principle is employed in integrated circuit FM detectors usually described as 'quadrature' or 'quadrature multiplier' (e.g. NSC LM1863, RCA CA3075). A small fixed external coil produces a 90° phase shift into an adjustable parallel tuned circuit centred on the modulated IF input. The S curve obtained from this single tuned circuit is linearized to give a typical 2% distortion for 1% deviation (e.g. 100 kHz in 10.7 MHz) by using a fraction of the rectified demodulated output to modulate the multiplier current. Likewise a small part of the rectified output controls a current source added to one side of the multiplier output. This compensates for any internal IC phase shifts at IF and tends to counteract DC offsets at the detector output over the RF tuning range. The tuned circuit produces a phase deviation relative to the instantaneous IF, of less than $\pi/2$, and the instantaneous IF can be the reference frequency. These two signals satisfy the required input conditions for a phase detector.

In a line-linked system a non-modulated 'drive' or comparison signal for the detector may be recovered from timing information contained in the received signal, but in a radio link this is not normally possible, hence the use of the single frequency-modulated signal as a source of both reference and phase-modulated inputs. It is important to realize that in this instance the recovered modulation is a measure of frequency deviation and not directly of phase modulation.

13 Private mobile radio

The cellular network is a public network that anyone with access to a telephone can use. Private mobile radio (PMR) on the other hand describes radio networks that are only accessed by a restricted group of users. A good example of this is taxi companies who use their radio to call cars for clients. Considerable use of PMR is made by police, fire and ambulance services.

Traditionally, private mobile radio has been used to cover a restricted area local to the transmitter. This is suitable for taxi companies covering a single town or district. An alternative is trunked PMR which allows calls to be routed to other transmitters, thus covering a larger area.

In order to ensure interoperability between different users, PMR is being standardized. Trans European Trunked RAdio (TETRA) is a Europe-wide system that is being introduced, primarily to allow different emergency services to talk to each other. Attempts are being made to make TETRA a worldwide standard, hence the proposed name change to TErrestrial Trunked RAdio.

13.1 Local private mobile radio

Local private mobile radio (PMR) normally employs a single-aerial site, with a number of aerials strategically arranged to cover the surrounding area. The effective range may vary between 10 and 50 km depending on aerial siting, power and frequency. Each station is licensed to operate on certain channels which can be allocated to specific vehicles. The radio station can be owned and operated by a public service or utility (police, ambulance service, water, gas, electricity, etc.) for its exclusive use, or by an operator who can hire out channels to local users with vehicle fleets. The office controlling the movements of the fleet is connected to the appropriate transmitter by means of a private line, and since the user has the exclusive use of certain channels, all communications with their fleet and between vehicles in their fleet are made through the office. This does not require any intervention by the site operator, allocation of channels on a call-by-call basis is unnecessary, and since the user has channels for exclusive use, no recording of calls for charge purposes is required, the user being free to make calls how and when he or she wishes. The system is therefore very cost effective for controlling a vehicle fleet within a defined local area.

Each channel will normally be shared between a number of vehicles with selective calling. The simplest system employs a range of audio frequencies allocated between the vehicles on a common channel. Alternatives are dual-tone multifrequency signalling (DTMF), continuous-tone controlled signalling (CTCSS) and various digital signals for transmitters having a digital capability.

Frequencies allocated for these services are in the VHF and UHF bands. Some bands use separate frequencies for send and receive, and some are single-frequency simplex.

All channels with separate base and mobile frequencies are two-frequency simplex. Simultaneous operation of base and mobile transmitters cannot occur, and 'press to talk' (PTT) operation is necessary.

Standard channel spacing is now 12.5 kHz which will accommodate double-sideband AM or low-deviation FM, for which the bandwidth requirements are $2f_m$ and $2(f_m + f_D)$ respectively (see Chapter 26). If FM is used therefore the deviation f_D must not exceed 2.5 kHz. The bandwidth will also permit digital transmission at 1200 bit/s using FFSK if required, but is insufficient for digitized speech.

Demands for PMR in relation to available bandwidth are high. Increasing the number of channels by reducing spacing could only be achieved by the additional complication of SSB working. Channel sharing between base stations is therefore essential (adequate distance being allowed between stations sharing channel frequencies to avoid interference). With channel sharing between base stations, and between mobiles using selective calling, a total of over 100 mobiles per channel is possible.

No provision is made in local PMR systems for access to the PSTN. The mobile handset can include a keypad with two-tone (MF4) dialling to provide direct switching access to other extensions on the switchboard.

The simple single-radio base station PMR installation is sometimes referred to as a 'community repeater', and will be seen to be considerably less complex than the multistation trunked PMR considered below.

13.2 Trunked PMR

PMR does not provide for the routing of incoming calls to the mobiles over a wide area with a number of distributed radio sites. This requirement adds a further stage of complexity. To cover a wide area involves a large number of base radio stations, and it is no longer possible to allocate specific channels to mobiles. To contact a mobile for an incoming call involves first locating the mobile to within its nearest radio station area, and then allocating a free channel on the appropriate base station. The setting up and monitoring of the call therefore has to be centrally controlled, and charging has to be on the basis of utilization; hence each call has to be recorded and charged.

To meet the requirement to cover a wide area, a nationwide network of PMR base stations is needed to replace the single station required for local PMR. These stations must be linked by land line, optical fibres or point-to-point radio, which may be leased or owned by the network operators. These systems are known as trunked PMR, and extend the scope of the local PMR to include large business organizations requiring a private network to maintain contact with their employees over a wide area.

The capital cost of setting up and operating a nationwide trunked PMR network is very high, and the number of networks that can be licensed is small. This has resulted in consortia being licensed to provide the networks.

Operators of PMR who own a number of radio base stations may interconnect these stations to form a trunked PMR and hence cover a wider service area. This provides a convenient means whereby a large organization or public utility can operate a wide-area private network. The networks must, however, comply with national standards, and therefore lose the simplicity of the local PMR.

In a multiple base station system contact between the mobile and its nearest base station has to be established. This is achieved by the base stations 'polling' the mobiles, using a fixed-frequency channel known as the forward control channel (FOCC). Frequencies for the FOCC are allocated to ensure that no adjacent stations have the same frequency, and each mobile can scan the complete range of FOCC to select the strongest signal, which will be that of the nearest transmitter. The 1200 bit/s bit stream in the FOCC base transmit channel indicates the time slots available for signalling in the associated mobile transmit FOCC, and also enables the mobile to synchronize with the base station. Conditions are therefore established which enable the mobile to make or receive calls via the nearest base station.

To make an outgoing call the mobile must seize the next vacant signalling slot, as instructed by the incoming FOCC bit stream, and transmit to the base station its own code and that of the required mobile or control office. The site controller in conjunction with the central controller routes the call to the appropriate base radio station and establishes contact with the required mobile or office. The mobiles are instructed to go to allocated free traffic channels and the call is established.

To receive an incoming call the mobile is paged via the incoming FOCC to which it is tuned, and responds on its transmit FOCC. On receipt of this acknowledgement the call can be established and the mobile receives an instruction to move to an allocated traffic channel.

The trunked PMR has no facilities for 'channel hopping' and a mobile must remain within the service area of the radio station through which the call has been established, for the duration of the call.

13.3 TETRA

TErrestrial Trunked RAdio (TETRA), formerly Trans European Trunked RAdio, is a digital private mobile radio system that allows interoperability between users. Because it is digital, speech can be encrypted before transmission to give increased security. This is particularly important for police and security services. Interoperability is also important because fire and ambulance services often need to talk to the police during emergencies.

The TETRA standard ETS 300394 has been defined by ETSI, the European Standards Organization. Frequency use is also being harmonized across Europe, with 380 to 400 MHz being allocated for 'public safety' use (i.e. the emergency services). Civil users are being allocated bands in the 400 to 500 MHz range and eventually the 800 MHz band.

TETRA uses Time Division Multiple Access (TDMA), allocating four users to share each frequency band of 25 kHz. The data stream comprises a start sequence, followed by 216 bits of scrambled data, a 52 bit training sequence, another 216 bits of scrambled data, and finally a stop sequence. The training sequence in the centre of the data stream allows the receiver to adjust its equalizer for optimum reception over the whole message. This is illustrated in *Figure 13.1*.

The data stream is time division multiplexed with three other digital channels. The radio carrier signal is then modulated using phase offset differential quadrature phase shift keying ($\pi/4$ DQPSK). The data stream is filtered by a root raised cosine channel filter in both the transmitter and receiver. This ensures an efficient use of the radio frequency spectrum by removing high order harmonics from the digital pulse. Overall raised cosine data filtering is achieved which allows

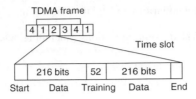

Figure 13.1 Digital data stream in TETRA

low-error signal recovery at the receiver. Transmit and receive frequencies are offset, allowing full duplex.

Speech is digitally coded using a voice coder/decoder (vocoder) designed by SGS-Thomson. This vocoder operates at the very low rate of 4.567 kbit/s. The vocoder generates codes for the speech it receives based on the human vocal system rather than the signal voltage at any instant in time. This prevents casual eavesdropping. The low data rate gives high spectral efficiency for the transmitted radio signal.

Data can be transmitted between radio sets. This can be either using a radio modem format or a packet data format (based on X.25). The data rate can be varied to suit the application; for example, transmission of a local map may require a high data rate.

Call set-up times for TETRA are short, about 300 ms. This is a feature of particular importance to the emergency services. A broadcast facility is provided so that, for example, all police vehicles can receive a message about an incident. Additionally, local one-to-one operation between TETRA radio sets allows users to operate outside the normal broadcast range; perhaps in a cave, or in the basement of a building.

13.3.1 TETRA radio

The TETRA radio transmitter has four power classes, capable of delivering a maximum of 1 W, 3 W, 10 W or 25 W. The power level is adjusted automatically according to the required signal strength.

13.4 Tetrapol

Tetrapol is a proprietary trunked radio originally developed by Matra in conjunction with the French Security Forces in 1987. The organization responsible for controlling the technical specification is the Technical Working Group (TWG) of the Tetrapol Forum.

Tetrapol has similarities to TETRA, including:

- Highly encrypted communications (including when using direct mode);
- direct mode communications;
- voice and data transmissions are supported (but not simultaneously);
- group and broadcast calls.

The fundamental difference is that Tetrapol uses FDMA (frequency division multiple access), whilst TETRA uses TDMA (time division multiple access) technology. Tetrapol supports one channel in either a 10 kHz or 12.5 kHz bandwidth (dependent upon the equipment manufacturer), whilst TETRA supports 4 channels in a 25 kHz bandwidth. Tetrapol supports data transmission rates up to 8 kbit/s.

13.5 Apco project 25

Apco Project 25 has been developed in the USA and is directed by the Association of Public Safety Communications Officials (APCO) and supported by the US Telecommunications Industry Association (TIA). Project 25 was created in 1990 with the objective of creating a common set of standards for digital trunked radio using Frequency Division Multiple Access (FDMA) technology. It also aimed to maximize radio spectrum efficiency and allow effective, efficient and reliable intra-agency communications.

Thirty-one standards were developed during Phase 1 of Project 25. The standards are based on six interfaces:

- Common Air Interface (CAI): ensures that the radio interface of all terminals is the same.
- Data Port Interface: terminals have a data port for connection of peripheral devices.
- Inter-System Interface: permits network infrastructure from different manufacturers to be used in a wide area network. It allows different technologies (TDMA, FDMA, etc.) and different frequency bands.
- Telephone Interconnect Interface: supports PSTN and ISDN.
- Network Management Interface: specifies a uniform network management interface.
- Host and Network Data Interface: supports TCP/IP, SNA, and X.25 connections.

It is possible that TETRA will be adopted in the next phase of Project 25 development. This is because TETRA uses TDMA, which offers greater spectrum efficiency, and because the TETRA standard is being adopted worldwide.

13.6 EDACS

Enhanced Digital Access Communications System (EDACS) was introduced into the European market by Ericsson GE Mobile Communications in March 1991. EDACS is a proprietary specification designed for the emergency services, offering integrated voice and data capabilities for the trunked radio user.

Features include:

- maximum data rate transfer of 9.2 kbit/s
- telephone interconnect
- Caller ID
- voice/message encryption.

It is unlikely that the standard will have a great impact, compared to TETRA, although by August 1999, Ericsson had more than 400 systems operating in both the public safety and commercial fields.

13.7 GSM ASCI/GSM-R

GSM ASCI (advanced speech call items) is an attempt to offer PMR-like features in a GSM network. It offers broadcast calls, group calls and priority calls. GSM ASCI features have been included in the GSM design for railway operator systems and is known as GSM-R. Although it is based on standard GSM infrastructure and engineering principles, GSM ASCI has been adapted to meet the following operational and safety requirements:

- Railway operations: man-machine interfaces.
- Railway telecommunications: broadcast and group calls, fast call set-up associated with the priority and pre-emption mechanisms.
- ERTMS (European Rail Traffic Management System) applications: this will later be extended to provide passenger multimedia services.

The GSM ASCI frequency band is just below the GSM-900 MHz band. This means that it requires its own separate infrastructure. GSM ASCI supports the new ETCS (European Train Control System) for the transmission of railway signalling information. GSM ASCI is not a threat to TETRA, because it is relatively expensive and is only intended for national and international railway systems.

13.8 GSM Pro

GSM Pro is a proprietary system developed by Ericsson. It offers PMR-like functionality from an upgraded GSM network and, as such, requires considerably less money to implement compared to building a TETRA network. GSM Pro suffers from the same limitations as standard GSM:

- long call set-up period (greater than 5 seconds)
- no prioritization
- low spectral efficiency
- standard demographic GSM coverage (as opposed to landmass coverage)
- standard capacity.

13.9 IDEN

Integrated Dispatch Enhanced Network is a proprietary system developed by Motorola. It uses the 800 MHz frequency band and TDMA technology. IDEN enables users to take advantage of wireless technology by integrating four communication services into one network, using one phone. That is, it offers business users advanced capabilities of dispatch radio (including group call, emergency priority call and direct mode), full-duplex telephone interconnect, short message service and circuit mode data transmission of 9.6 kbps.

IDEN is aimed primarily at the business user in the USA and is not envisaged to become widespread in Europe, since other European PMR technology provides greater functionality. Also, IDEN is a proprietary technology, like Tetrapol, and therefore will not satisfy customers that require systems based on an open standard.

14 Conversion between analogue and digital

Analogue to digital conversion is required where analogue signals (i.e. speech and modem carriers) need to be carried by digital transmission systems. The simplest type compare a sample of an analogue signal with a range of predetermined voltage levels, then output a digital code that represents a voltage nearest to that of the sample. More complicated systems reduce the data rate by producing an output dependent upon the difference between one sample and the next. Vocoders are even more complex, and only suitable for speech, producing an output representing a vocal sound.

Conversion from digital back to analogue is needed at the end of the digital transmission system, where analogue signals are locally transmitted to customers. Generally, digital to analogue converters are much simpler than analogue to digital converters.

14.1 Analogue to digital conversion

14.1.1 Pulse code modulation (PCM)

Pulse code modulation is achieved in two stages: (1) the analogue signal is sampled at 8 kHz; and (2) the sampled signal is converted into a digital word of 8 bits. The 8 kHz sample rate limits the analogue frequency to 3.4 kHz. The lower analogue frequency limit is 300 Hz, due to the performance of coupling capacitors and transformers prior to sampling.

In PCM the magnitude of each sample with respect to a fixed reference is quantized and converted by coding to a digital signal. Quantizing involves rounding off the instantaneous sample pulse amplitude to the nearest one of a number of adjacent voltage levels. *Figure 14.1* illustrates the process for an eight-level system. In the figure, the analogue amplitude is 2 at $t = 0$, between 4 and 5 at t_1, between 5 and 6 at t_2, etc.

The quantized levels would be 2 at $t = 0$, 5 at t_1 and again 5 at t_2 and so on, taking the nearest rounded-off value at each sample. The difference between analogue and quantized signal is quantization noise or distortion.

166

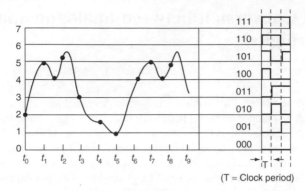

Figure 14.1

The binary pulse train would therefore be (leaving a 1 bit synchro-nizing space between each number) as shown in *Figure 14.2*. Note that the pulse amplitude is constant but the pulse pattern changes according to the quantized analogue signal.

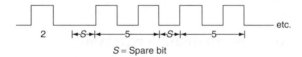

Figure 14.2

The number of quantizing levels is 2^n and the highest denary-numbered level is $(2^n - 1)$ where n is the number of bits used to represent each sample. If each train of pulses representing a sample is accompanied by one synchronizing bit, the total number of bits per sample is $(n + 1)$. If the sampling rate is f_c, the transmitted bit rate is $(n + 1)f_c$, and the maximum fundamental frequency of the pulse train will be when alternate '1' and '0' pulses occur. Hence the max-imum transmitted frequency is $\frac{1}{2}$(bit rate). The minimum transmitted frequency, with unipolar pulses, will be 0 Hz when consecutive '0' and '1' occur.

In the above example $n = 3$, but all PCM systems use $n = 8$ (1 sign bit and 7 data bits), so there are 256 quantization levels. One should note that the 24 channel PCM system (T1) used in the USA and Japan employs 'bit stealing'. Bit stealing occurs once every six PCM frames where the least significant bit of all channels is stolen and used for signalling. This is another reason why the so-called 56 K modems ignore the least significant bit. In Europe, bit stealing does not occur in the 32 channel PCM system (E1) because a dedicated signalling channel is provided (time slot 16). Only 30 channels are available to customers because time slot 0 is used for synchronization (the time slots are numbered 0–31). The T1 system also uses synchronizing channels, and so allows 23 channels to be used by customers.

Linear quantization, as illustrated above, would produce a poor signal/noise ratio for low-level signals, and so a non-linear characteristic is employed in which there are more closely spaced quantization levels at small-signal amplitude than there are at maximum signal amplitude. The two most commonly used non-linear transfer characteristics are 'A' law and $\mu 255$ law, the former being the ITU-T recommendation used by the European E1 system and the latter being commonly used by the T1 system in the USA and Japan. The characteristics are similar, both being a piecewise linear approximation to a quasi-logarithmic continuous function, and both allowing for quantization of positive and negative excursions of the analogue signal from reference zero. The general shape of the characteristics is given in *Figure 14.3* from which it can be seen that the most significant bit of the 8 bit sample character determines the sign of the analogue signal. At the receiving station, the decoder will have the complementary characteristic so that the overall analogue/analogue transfer is linear. The process is called COMPANDING – COMPression and exPANsion – and the circuit referred to as a CODEC – CODe/DEcode.

Modern LSI has made it possible for the whole sampling, PCM encoding/decoding process and filtering to be carried out in one

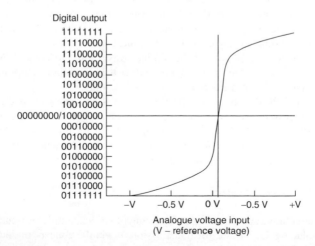

Figure 14.3

integrated circuit. For example, the Mitel MT8960 series announced in 1982 provide a complete conversion interface between analogue signals in a telephone subscriber loop and the digital signals required for a PCM system. Analogue signals entering the chip are first bandwidth limited by switched capacitor filters to 3.4 kHz, and also have any 50/60 Hz component removed by a high-pass notch filter. Any accumulated DC offset is removed by an auto-zero loop. Typical filter characteristics (relative to 800 Hz) are 30 dB attenuation for 0–60 Hz and 35 dB attenuation at 4.6 kHz and above. The digital encoder generates an 8 bit word in accordance with 'A' law or 'μ' law and this is output at a nominal rate of 2.048 MHz as the first 8 bits of the 125 µs (i.e. 1/8000) sampling frame.

14.1.2 Delta modulation

Delta modulation is used because less hardware is required, and hence the cost is lower. If the speech waveform's instantaneously sampled value is greater than the previous sample a binary 1 is transmitted; if it is equal to or less than the previous sample a binary 0 is transmitted. The sampling rate is typically 32 kHz, but each binary sample signal is either 0 or 1 (compared with an 8 bit PCM 'word'). Thus the data rate is half that of a conventional PCM analogue to digital converter.

14.1.3 Differential pulse code modulation

A speech waveform is predictable to some extent, because the rate of change is slow compared to the maximum rate allowed by sampling theory. One sample can be predicted from knowledge of previous samples. Although errors will occur, these will be far smaller than the maximum amplitude of the original waveform. Differential PCM transmits the error in the prediction, rather than the sample amplitude, and can thus transmit fewer bits for each sample.

The DPCM converter typically uses a 4-bit analogue to digital converter to convert the input signal into an output code. The digital output is converted back to an analogue signal for use in a feedback circuit. A delayed output signal is fed back for subtraction from the input. This is illustrated in *Figure 14.4*.

14.1.4 Speech codecs

An alternative approach to the digitizing of speech, aimed particularly at reducing the bit rate for transmission, is via the concept of speech synthesis.

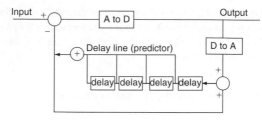

Figure 14.4 Differential PCM converter

Speech waveform characteristics (of a particular language) such as vowel sounds, consonants, nasals and fricatives can be expressed in terms of the coefficients of the filters used to extract them. The digitized filter coefficients extracted from an input speech signal are transmitted, together with an 'excitation' sequence at a relatively low data rate; for example the Pan-European (GSM) digital cellular network uses data transmission at 13 kbit/s.

14.1.5 Video digitizers

Improvements in technology and increasing demands on allocation of the radio-frequency spectrum are making the transmission of video signals by wire an attractive and realizable proposition.

A composite PAL colour signal requires a bandwidth in excess of 5.5 MHz and 8 bit linear resolution. The usual sample rate is 13.3 MHz (i.e. three times the colour subcarrier) which for 8 bit resolution gives a bit rate of 106 MHz. Use can be made of differential PCM to reduce the bit rate; an estimate is made of the next sample based on preceding samples and the difference between the predicted and actual next sample is transmitted with 5 bit resolution (i.e. with only 32 PCM quantizing levels instead of 256). The eye is relatively insensitive to small changes in a level of high luminosity and a further bandwidth reduction is possible using tapered rather than linear quantization (cf. 'A' law speech quantization).

In practice a 66.5 Mbit/s rate has been used, multiplexed with a 2.048 Mbit/s channel for high-quality sound plus synchronization, etc., to give a total bit rate of 68.736 Mbit/s for transmission on the 140 Mbit system.

ITU-T Recommendation H120 is the standard for existing video conferencing codecs; video conferencing pictures are generally more static than broadcast TV and codecs designed to H120 use a technique called 'conditional replenishment' in which only those parts of an image which change from one frame to the next are encoded by

DPCM. An update, H261, for the next generation of video conference and videophone codecs will apply to bit rates in multiples of 64 kbit/s up to 1920 kbit/s, and will cover compatibility with 625 line and 525 line standards of television.

14.2 Digital to analogue conversion

Conversion from digital to analogue is quite simple for an 8-bit binary word, such as used in PCM systems. The digital code is output in parallel form, each output passing through a series resistor. The resistor values are weighted such that the most significant bit outputs have the lowest resistor values connected. The resistors are commoned together and taken to an analogue summing circuit. As the digital code changes to represent higher input voltages at the analogue to digital converter, the output of the summing node follows those changes.

14.2.1 Differential PCM converters

At the analogue to digital converter, a predictor estimates the output for the next sample. Only a digital code representing the error of the predictor is transmitted, therefore the digital to analogue converter at the receiver must also have a predictor. The transmitted code is then used to correct the predictor output at the receiver, so that the system output is a close representation of the input.

15 Line interface functions

Copper pairs used in the local telecommunications network are permanently connected between fixed points, since they are not robust enough to be continually connected and disconnected. At the customer's end the wires are terminated in a telephone socket (typically). The standard telephone socket provides limited functions, but ISDN terminations are more sophisticated. These interfaces are described further in Section 15.1.

At the telephone exchange (central office) end the wires are terminated on a distribution frame, which provides a point where cross-connections can be made and replaced many times. The distribution frame has two sets of termination points, one set for the permanent external cable termination, and another for the permanent connection to interface circuits. Temporary copper pairs, called jumper wires, are used to join pairs on the external cable terminations with pairs on the exchange's interface circuit terminations. The exchange's interface circuits provide many functions and are described in detail in Section 15.2.

15.1 Interface functions at the customer's end

Screw terminals or insulation displacement connectors provide a fixed termination for the exchange line wires. There are only two wires from the exchange, but up to six connector terminals are provided. Some terminals are for use when extension telephones are required, one is for use by PBX extensions for connecting a recall earth (an earth applied to the line signals the PBX that the user wants to transfer the call to another extension).

Customer equipment such as a telephone, facsimile machine or modem can also be connected via screw terminals, to give a fixed connection between the equipment and the line. More often customer equipment is connected via a plug and socket arrangement; this allows equipment to be moved or changed by the customer, as required. The most common plugs are rectangular; for example, RJ11 in the USA (where the exchange line is on the middle pair of pins (3 and 4)) and Type 431A in the UK (where the exchange line is on pins 2 and 5).

There are many variations of telephone socket in the world. Most have just an exchange line connection and a telephone socket. The UK is unusual in that the bell capacitor, a resistor for line testing and a gas discharge tube are provided in the telephone socket. This allows the network operator to protect and test the network, without relying upon the customer's telephone to provide these functions. *Figure 15.1* shows the UK socket; more details on this are included in Chapter 26.

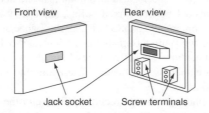

Front view   Rear view

Jack socket   Screw terminals

Figure 15.1   A typical telephone socket

ISDN terminals have the exchange line terminations and equipment sockets. They also have line transmitter, receiver, multiplexing and clocking circuits. Power is provided locally, but power feeding over the exchange line is possible. The ISDN terminal transmits and receives data continually, even with no terminal equipment plugged in. The ISDN terminal and the exchange clocks remain synchronized at all times; if the line is broken the system locks up, causing an alarm in the exchange. A maintenance technician or engineer has to reset the system in the exchange after such a fault. These terminals are large, compared to normal telephone sockets, typically 100 mm wide by 250 mm high and perhaps 30 mm deep.

15.2 Interface functions at the telephone exchange (central office)

There are several functions to be provided by the telephone exchange line interface, these are known by the initials BORSCHT: Battery feed, Overvoltage protection, Ringing, Signalling, Coding, Hybrid and Test.

15.2.1 Battery feed

This is the exchange battery current that flows when the telephone handset is lifted. The purpose of the battery voltage is three-fold: to allow the telephone to signal the exchange when the handset is lifted; to allow the customer to signal the exchange a number to be called; and to provide power to the microphone from which the speech signals are generated. In older telephone exchanges battery feeds are nominally 48 V fed through two windings of a relay coil, each having a resistance of 200 Ω. The power feed arrangement allows the relay to act as a transformer. It is chosen to have high inductance, thus presenting a high impedance in parallel to the load at audio frequencies. A secondary winding is provided to enable the a.c. signal to be switched through the exchange whilst restricting the d.c. path to just the telephone loop. The relay contacts close when the telephone handset is lifted, due to the d.c. current flow. The contacts are used to operate switching circuits in the exchange, either directly or via a processor (simple computer).

The copper wire line to the customer can have a loop resistance of up to 1000 Ω. The telephone instrument itself has in-built regulators that attempt to keep the line current reasonably constant; a resistance of between 500 and 1000 Ω can be expected when the handset is lifted. The lower resistance is more likely on longer lines. The total load can therefore vary between about 1400 and 1900 Ω, giving line currents of between 35 and 26 mA respectively. At the telephone instrument, with the handset lifted, the line voltage will be about 35 V on short lines, but could be as low as 15 V on long lines.

The telephone produces an audio frequency signal with a power level of less than 1 mW. However the d.c. power reaching the instrument is between 1.225 W and 338 mW; not very efficient. Remember though that this power is also used to signal the exchange that the handset is lifted and for dialling a number for an outgoing call.

Some telephone exchanges (particularly digital ones) have a different battery feed arrangement. A constant current feed is provided to overcome the problem of having a telephone terminal voltage dependent upon the line length. The nominal current is 30 mA and has an open circuit voltage of 72 V. This combination allows the copper wire's loop resistance to be as high as 2 KΩ, almost twice that of a constant voltage feed. A graphical comparison between the different power feed arrangements is given in *Figure 15.2*.

The resistive, constant voltage feed produces a decaying terminal voltage as the line resistance increases. Typically, the working voltage of terminal equipment is 15 V or more. Thus a line resistance of up to 1.2 kΩ can be used.

The constant current feed is produced from a higher voltage source. The terminal voltage remains at about 25 V. As the line resistance increases, so does the voltage dropped along it. At a line resistance of about 1.6 kΩ, the total voltage drop of the line and load equals 72 V. This is equal to the source voltage and is thus the limit of constant current working. Lines with a resistance of more than 1.6 kΩ do not

174

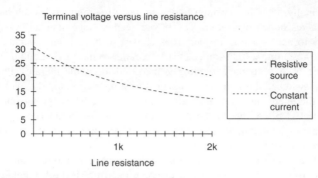

Figure 15.2 Exchange power feeding

have a constant terminal voltage. Terminal equipment will continue to work over line resistances of 2 kΩ or more.

15.2.2 Over-voltage protection

Lightning can induce voltages into telephone cables that would cause damage to the exchange equipment. Some form of discharge device is required to prevent this. Gas discharge tubes are sometimes used; these have an earthed electrode in between two electrodes connected across the copper pair. When a high voltage is induced into the line, the gas ionizes (breaks down and conducts), to safely short circuit the induced voltage to earth. The gas discharge tube returns to its insulating state once the high voltage has been removed. The ionization voltage of gas discharge tubes varies, but devices that have a breakdown voltage of under 250 V are used.

An alternative to the gas discharge tube is a semiconductor 'varister' that breaks down when high voltages are applied, but which restores to an insulating state after the line state returns to normal. The advantage of the varister is that it reacts faster than the ionizing gas discharge tube, thus giving more protection.

15.2.3 Ringing

Ringing is required to let the customer know that someone is trying to call them. The ringing signal has a frequency of about 20 Hz and a typical amplitude of 75 V rms; these values vary from country to country. Ringing is applied in bursts, rather than continuously. The pattern of ringing bursts is called cadence. In the UK, the cadence of ringing is two short bursts with a short pause between, followed by a longer pause. Other countries use telephone systems with a different cadence. In the USA, for example, the cadence has one burst of ringing followed by a long pause.

15.2.4 Signalling (or supervision)

This is monitoring the line conditions and signalling this to the exchange processor. When the handset is lifted the exchange must apply dial tone and prepare to decode dial pulses or DTMF tones. Dialled numbers must be interpreted for the exchange processor so that it can route the call as necessary. The call must be cleared when the handset is replaced.

On incoming calls the exchange must know when the handset is lifted. When this condition is detected the ringing generator is turned off, the ringing tone removed from the caller's line, and the call is switched through so that the two parties can talk.

15.2.5 Coding

In digital exchanges, analogue to digital and digital to analogue conversion is required. These conversion processes were described in Chapter 14.

The digital word, or byte, is 8 bits long and could be transmitted in series or parallel. Each bit is a logic state, 1 or 0; a logic 1 is a positive voltage, typically $+5$ V and a logic 0 is typically 0 V. The pattern of logic states in the 8 bits can be made to represent an analogue voltage, thus 256 analogue levels can be represented, although half are positive and half negative. The most significant bit represents the analogue signal's polarity.

15.2.6 Hybrid

The transmit and receive signals are both present on the line between the customer and the exchange. Analogue switches within an exchange can also pass bi-directional signals. In circuits where signal process takes place, such as in amplifiers, analogue multiplexers or digital switches, we require the transmit and receive signals to be separated. A hybrid is a device that allows the separation of transmit and receive signals.

Older analogue trunk circuits are usually configured for 4-wire (i.e. two pair) transmission. One pair is used for transmission from A to B, whilst the other is used for transmission from B to A. This allows the signals to be amplified easily, but conversion from 2-wire to 4-wire transmission is required at either end of the trunk circuit. This conversion can accomplished with a passive hybrid circuit, each using two transformers, as shown in *Figure 15.3*. Active hybrids, using operational amplifiers, are popular because of their small size and are generally replacing passive devices.

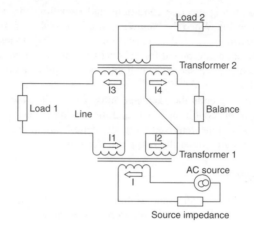

Figure 15.3   Hybrid 2–4 W conversion

Current from the source flows through the primary winding of Transformer 1. This generates currents in each secondary winding. Current I1 flows through one winding Transformer 2 (labelled I3) and through Load 1. Current I2 flowing in the other secondary winding of Transformer 1 also flows through the balance impedance and a winding in Transformer 2. Currents I3 and I4 flow in opposite directions through the windings of Transformer 2, so magnetic flux produced in the core is due to the difference between these currents. Current flow I1 (=I3) depends upon the impedance of Load 1 and current flow I2 (=I4) depends upon the balance impedance. For zero current through Load 2, the impedance of Load 1 should be equal to the balance impedance.

15.2.7 Test

Access points are required in the line interface circuit so that the line can be tested. It is also necessary to test the interface circuit functions. Line tests include: loop resistance; insulation between the wires of a pair; and bell circuit capacitance (usually detected by reversing the polarity of a d.c. test voltage and observing the 'kick' of the meter's needle). Exchange equipment tests include: ringing current; exchange supply voltage; dial tone detection; and dial pulse or MF4 tone decoding checks.

15.3 Unbundled local loop (ULL)

The US, German and Australian governments have introduced regulations requiring incumbent local exchange carriers (ILECs) to make their copper access network available to other companies for the purpose of wide-band access. Wide-band access systems, such as ADSL, were described in Chapter 8. The EU issued a regulation that required all ILECs in member states to publish terms for access to the copper access network by 31 December 2000. These terms had to include non-discriminatory access with equal terms and conditions.

The EU regulation includes sub-loop access, local loop unbundling, line sharing and bit-stream access. In Germany, unbundling has been underway since 1996 and access to fibres also comes under the local regulations. Germany has the largest customer base (some 50 million lines) and so far there are 300 000 unbundled local loops. The unbundled lines are only in major cities and represent an overall 0.6%. However, ISDN is very popular in Germany, which gives many Germans moderately fast Internet access already.

15.3.1 Local loop unbundling

Local loop unbundling is when other telecommunication service providers can have complete access to the copper pair, providing data alone or telephony and data services to their customer over existing lines. There are two possible configurations. One configuration is the provision of a second copper pair to the customer. The first copper pair continues to carry telephony services from the ILEC. The second copper pair is used to carry data from the alternative carrier. This configuration is shown in *Figure 15.4*.

In the alternative unbundling configuration, the existing copper pair is transferred to the alternative carrier. This company then provides telephony and data services. The ILEC no longer provides services to the customer, but maintains the line under contract from the alternative carrier. This configuration is shown in *Figure 15.5*.

15.3.2 Line sharing

Line sharing is when the alternative carrier only wants to provide a data service and the customer keeps his telephony service with the

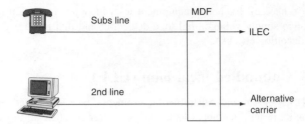

Figure 15.4 Unbundling with second copper pair

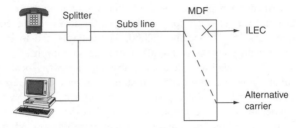

Figure 15.5 Unbundling by transferring existing pair

ILEC. This is likely to be very popular with internet service providers (ISPs) and entertainment companies, who are not very interested in providing a lot of new infrastructure in order to carry telephony. In this situation the ILEC continues to provide telephony services, either directly or via the ISP who acts as a virtual telephone company.

In some cases the alternative carrier provides the 'splitter' used to separate the telephony from the ADSL, but regulators are considering proposals for the ILEC to provide this. The proposed configuration is shown in *Figure 15.6*.

15.3.3 Bit-stream access

Bit-stream access is when the ILEC provides all the line terminating equipment, such as ADSL, but allows another operator to transport data through it and onto the customer. This is attractive to some companies who do not want to get involved with the engineering aspects of telecommunications.

15.3.4 Sub-loop access

The sub-loop access condition in the EU regulations means that the ILEC must allow competitors access to the local loop, at any point

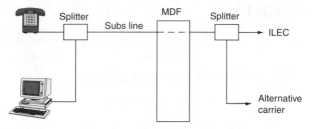

Figure 15.6 Line Sharing with ILEC Provided Splitter

between the main distribution frame (MDF) in the local exchange and the customer premises. The purpose of this regulation is to allow the future use of very high-speed digital subscriber line (VDSL) equipment. VDSL provides access data rates of up to 55 Mbit/s, but this has limited reach. The intention is to provide high-speed data over fibre to a point close to the customer (maybe the street cabinet) and then piggyback the service over the copper pair into the customer's premises.

15.3.5 Local exchange access

All ILECs are expected to provide accommodation for competitors 'where space permits'. This has caused a few problems in defining whether space is available. The incumbents can charge for converting the rooms and charge a rent for the space used.

Some ILECs provide a room with large cages in which alternative carriers can place their equipment. Mains supply, air conditioning and tie cables to the MDF are provided. In the USA, alternative carriers are allowed to install their line terminating equipment (DSLAM) alongside the ILEC.

16 Switching

16.1 Electromechanical components

16.1.1 Relays (conventional armature)

Electromechanical relays are still extensively used in telecommunications to fulfil a multiplicity of functions:

(a) nominally simultaneous operation of a number of switching contacts;
(b) insulation of control and switched circuits from each other;
(c) power amplification;
(d) separation of AC and DC circuits (i.e. DC control of AC power);
(e) delaying or shaping a control signal.

Modern relays are in a variety of forms ranging in size from the subminiature in a T0.5 transistor package to heavy-duty types on 8 or 11 pin plug bases or even in chassis-mounting bolt-down form. Parameters to be considered in the selection of a relay are:

1 *Coil voltage and current*: Separate consideration must be given to the 'pull-in' voltage at which the relay just operates and the 'drop-out' voltage which is the maximum level guaranteed to restore the armature to its rest (zero control) position. There will also be a maximum voltage limitation due to permissible power dissipation in the relay coil. Thus a relay catalogued for 12 V operation may have a drop-out level of 3 V, a minimum pull-in level of 9.5 V and a maximum rating of 18 V.
2 *Contact material and rating*: Separate voltage, current and volt-amp ratings are quoted dependent on the type of service use. For example, a maximum switched voltage rating may be 150 V DC but 120 V AC; maximum carry current may be 2 amps but maximum switched current 1.25 amps AC or DC and, for the same contacts, the maximum switched power may be 50 VA. Silver/nickel oxide or silver/cadmium oxide contacts are common for medium to high currents (0.5–10 amps). For so-called 'dry' contacts where the switched voltage and current are of the order of 100 mV and a few microamps the use of gold-flashed (i.e. 0.1–10 μm thickness) contacts is essential to avoid problems with contact resistance. Rhodium plating, though expensive, is sometimes used

as a hard-wearing surface in reed relays. Contacts are sometimes referred to as form 'A' – normally open; form 'B' – normally closed; or form 'C' – changeover. In high-frequency situations, for example in a relay-operated wideband attenuator, the capacitance between open contacts, between contact sets, and from contacts to ground may assume importance.

3 *Operate and release times*: The total time delay between application of a DC control voltage and closure of a contact consists of the delay in armature movement as the DC current builds up exponentially in the coil inductance, plus the travel time of the armature. Depending on the type of relay and coil, this could be in the order of 2–20 ms.

Likewise, when the control voltage is removed, there will be a similar though probably shorter time delay before contacts return to the rest position.

The drop-out time of a relay can be artificially increased by a number of methods, and is often done in order to achieve a required sequence of operations.

(a) A delay up to a few 100 ms can be achieved by short-circuiting an auxiliary winding on the same core via a normally open contact.

(b) Delays in the order of milliseconds can be achieved by permanently short-circuiting an auxiliary winding via a diode.

(c) A suitable capacitor placed across the relay coil can provide delays up to about 20 seconds since, when the supply voltage is switched off, the capacitor discharges through the coil resistance and the relay remains operated until the capacitor current falls below the relay hold-in current. A resistor may be required in series with the capacitor.

4 *Bounce*: This occurs mostly on closure of a pair of contacts and can cause contact interruptions for a period of some milliseconds. Even after contact interruptions have ceased there can be a further interval of contact movement causing modulation of the contact resistance. These effects can cause malfunction of quick-acting semiconductor switch and logic circuits and a semiconductor latch or 'bounce eliminator' should always be used to interface a mechanical contact set to a logic circuit (*Figure 16.1*).

5 *Life*: This is affected by switched current and voltage, the nature of the switched load (resistive, reactive, high in-rush), contact bounce, operating environment, frequency of operation, etc. It can be quoted as:

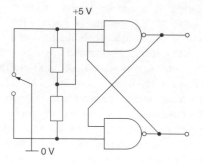

Figure 16.1

(a) Electrical life: the number of switching cycles which, with a
specified contact load and operating voltage, is permissible with
a specified (e.g. 95%) probability of survival. A figure of 10^5
would be typical.

(b) Mechanical life: the number of switching cycles (50% on-off)
with no contact load before failure to meet criteria of pull-in/
drop-out voltage, or of insulation. A typical figure could be
5×10^6.

16.1.2 Latching (bistable) relays

These retain the switched position after interruption of the control
current, either mechanically or by means of a permanent magnet.
Two windings may be used where only one polarity of control signal
is available to set and reset the relay. If bipolar control is available
only one winding may be required.

Since latching relays can be pulse operated their use is advantageous to power saving in, for example, battery-operated equipment.

16.1.3 Reed relays

In dry reed relays the armature and contact set are combined into
one simple cantilever assembly using ferromagnetic material. In its
basic form A style, this assembly is sealed into a glass tube and can
be actuated by a concentric coil carrying the control signal, or, as in
some process control systems, by the approach of a small permanent
magnet.

Because of their small inertia, dry reed contacts have an operate
time of 1–2 ms including bounce. Also, their low inductance
and capacitance make them ideal for RF applications. Typical

ampere-turns to operate range from 20 to 80 for form A and up to 100 for form C contacts. Maximum contact ratings are typically 0.25–2 amps, 5–20 V A with a mechanical life of 10^8 operations.

Reed relays with operating coil are also available in plastic dual-in-line packages on a standard 0.1 inch grid. Reed relays form the basic switching element in a crosspoint switch matrix. Analogue signals are switched under the control of a computer.

16.1.4 Mercury-wetted reed relays

In these relays the glass tube containing the contact also has a pool of mercury to keep the contacts wetted and may contain an inert gas under pressure [USER BEWARE]. This gives bounce-free switching with low and constant contact resistance, but with the disadvantage that the relay must be used within $\pm10°$ to $\pm15°$ of vertical. Typical electrical life at 24 V 2 A is greater than 50×10^6 operations.

This relay can be operated up to about 100 times per second and has been used to charge and discharge a length of coaxial cable, producing high-level (e.g. 50 V) pulses with rise and fall times of the order of 1–2 ns, and of duration determined by the electrical length of the cable.

16.1.5 Contact protection

Relay contacts must be protected from the deleterious effects of arcing which can occur when a contact pair opens, particularly if there is inductance associated with the switched load. Even with a purely resistive load, arcing can occur if the voltage being switched is greater than the critical voltage of the gap between the opening contacts.

16.2 Telephone exchange (central office) switching

16.2.1 Numbering and call routing

Every terminal station requires an individual numerical identity. From the user's point of view a decimal number is the most convenient and has been accepted as the universal standard. This fitted in well with the electromechanical automatic switching systems, where each digit formed a step in the call routing and the digits were transmitted as make/break impulses at a rate of 10 impulses per second.

Modern electronic switching systems separate the actual switching from the control of the switching, and the actual routing information is conveyed as binary signals in a high-speed bit stream. The process of conversion from a decimal to a binary number is now such a routine function that the decimal numbering for terminal addresses has been retained, and both impulse dialling and keysending operate on this basis. The numbering convention is normally 1, 2, . . . , 8, 9, 0, the '0' representing a count of 10.

Each terminal is connected to a local exchange and has its individual number within the exchange numbering scheme. Depending on the number of terminals and exchange capacity, this number can require between three and seven digits. When dialling a number on the local exchange, only these digits need be dialled, and only switching in the local exchange is required to route the call. Additional switching, and consequently additional numbering, is required to route calls first to other local exchanges, then to exchanges in the same national network, and finally to other countries.

The evolution of satisfactory numbering systems has been complicated by other factors:

(a) The increasing demand for 'direct dialling in' (DDI) to PABXs. This requires additional digits, giving rise to a 'multiple subscriber number' (MSN) which could require up to eight digits.
(b) The possibility of more than one PTO (Public Telecommunications Operator) sharing a national numbering scheme.
(c) The fact that in addition to the PSTN, the PTOs operate a number of other networks such as telex, PSDN and packet switching, requiring their own separate numbering schemes.
(d) Numbering schemes must also cater for any of the mobile radio systems which utilize the facilities of the PTOs.

To meet these requirements ITU-T have drawn up a series of recommendations, which are being adopted internationally and are listed in Chapter 26. Some typical address structures, are shown in *Figure 16.2*. Two formats are shown for the PSTN. ITU-T E163 (*Figure 16.2(a)*) was introduced to embrace older equipment which requires the code to carry the actual sequential switching instructions. This is a logical system but results in different dialling codes dependent on the location of the calling station. It also limits a call routing to one route.

The widespread adoption of stored program control in digital exchanges has enabled the routing sequence to differ from the dialled number, the routing sequence corresponding to any dialled routing

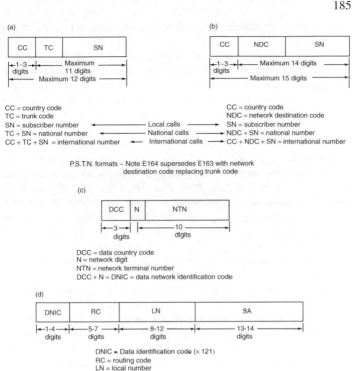

(a)

CC	TC	SN

|← 1–3 digits →|← Maximum 11 digits →|
|← Maximum 12 digits →|

(b)

CC	NDC	SN

|← 1–3 digits →|← Maximum 14 digits →|
|← Maximum 15 digits →|

CC = country code
TC = trunk code
SN = subscriber number
TC + SN = national number
CC + TC + SN = international number

◄——— Local calls ———
◄——— National calls ———►
◄——— International calls ———►

CC = country code
NDC = network destination code
SN = subscriber number
NDC + SN = national number
CC + NDC + SN = international number

P.S.T.N. formats – Note E164 supersedes E163 with network
destination code replacing trunk code

(c)

DCC	N	NTN

|← 3 digits →|← 10 digits →|

DCC = data country code
N = network digit
NTN = network terminal number
DCC + N = DNIC = data network identification code

(d)

DNIC	RC	LN	SA

|← 1–4 digits →|← 5–7 digits →|← 8–12 digits →|← 13–14 digits →|

DNIC = Data identification code (× 121)
RC = routing code
LN = local number
SA = sub-address

Figure 16.2 ITU-T address structures, (a) ITU-T E163. (b) ITU-T E164. For PSTN formats, note that E164 supersedes E163 with network destination code replacing trunk code. (c) Public switched data network (PSDN) – X21. (d) Packet switchstream

numbers being stored in the exchange processor. Such systems are referred to as director systems, and have the advantage of enabling a standard national dialling code (NDC) to be used for each terminal, and also of permitting alternative routes to be programmed and used if the first choice route has no free path. The address format standardized for these systems is shown in *Figure 16.2(b)*.

To implement the routing process the country (or territory covered by the PTO) is split up into local areas, each of which is allocated a three-digit code, permitting a total of 999 areas. The areas act as the first concentration point for calls, and also as charge areas, the called area number determining the charge rate according to the location of the caller. In the USA, the area three-digit number is preceded by a one, providing a four-digit NDC (*Figure 16.2(b)*). The one is not required on international calls and is therefore dropped for incoming international calls.

The area may have a single exchange, an exchange with satellite unit automatic exchanges, or several exchanges. Routing to alternative exchanges in a local area must involve additional digits, and when required, these digits must form the first digits in the SN sector. This enables a group switching centre, associated with the local area, to route calls to any exchange in the area, to adjacent areas, or to a trunk main switching centre. Trunk main switching centres are normally located in cities, and accept trunk calls from the group switching areas allocated to their region. Heavy-traffic concentration exists between main switching centres, and fibre optic links are normally used for these. If a call is outside the local area, the NDC will be included, and the '1' at the beginning of the NDC indicates that the call is not local. The NDC indicates the local area to which the call is to be routed, but the actual digits required for the routing (and alternative routing) are called up from a look-up table in the switching computer's memory.

At each stage of routing, the digits required for the following stages must be transmitted on. In analogue systems these are transmitted as part of the channel signal, and may take the form of impulse interruption of a DC loop, in-band, or out-of-band tone signalling. In the case of digital signals, a time slot (or bits stolen from a number of time slots) in a multiplexed data stream, is provided to carry the routing signal.

16.2.2 Electromechanical switching

Electromechanical switching equipment, using single- and two-motion electromagnetic switches, was the first successful automatic system to be introduced, and is normally referred to as the Strowger system after the inventor. These systems are now obsolete, but have formed the backbone of the automatic network over a period of 50 years. Although by modern standards the Strowger techniques are bulky, noisy and slow, the system architecture is simple and elegant, and a short description provides a useful introduction to automatic switching.

The uniselector has banks of 25 contacts arranged in a 180° arc, up to eight banks per switch being available. A wiper assembly, with one wiper per bank, is rotated by means of a magnet, armature, ratchet and pawl mechanism. The uniselector therefore provides selection of one in 25 outlets. If half of the wipers are assembled with a displacement of 180°, the uniselector provides a selection of one in 50 outlets with a 360° rotation of the wiper assembly. Uniselectors are used as line selectors and in the relay logic circuitry associated with switching.

Two-motion selectors have up to three banks, each having 100 contacts, arranged in 10 levels with 10 contacts per level. One wiper per bank is fixed to a vertical shaft which has vertical and horizontal magnet and armature drives. These operate via ratchet and pawl mechanisms first to drive the wipers to the correct vertical level $(1, 2, \ldots, 0)$, and then to the correct outlet in the level. Outlet 6 on level 4 would represent outlet contact 46.

At the conclusion of its operating cycle, each type of selector must return to its ready position. With a uniselector this is achieved by self-driving round to the 'home' position, and with a two-motion selector, either by means of a release magnet or by self-driving until the wiper clears the row, and then dropping down and across to the start position. The magnetic drive systems of both types of selector are designed to operate with the 10 ips dialling signals, with a 1:2 make/break ratio.

With the Strowger system, route switching employs the two-motion selector almost exclusively, the 10 levels with 10 outlets per level enabling direct switching to be achieved with a decimal dialling sequence. Economic design dictates a sharing of two-motion selectors, the provision of switching capacity being determined by maximum traffic density. Only the final selector in the switching sequence can therefore use both vertical and horizontal motions for address digits; all earlier selectors use the vertical motion to switch one digit and the horizontal motion to select a free selector for the next digit. Similarly a two-motion selector cannot be provided for each terminal, and concentration has to be provided either by a uniselector associated with each line, which selects a free first selector, or by a reverse process by which the first selector finds a calling line (line finder). The route switching is therefore as shown in *Figure 16.3* which includes a typical routing to subscriber 3645, and also the way in which the routing digits are used to switch to other exchanges.

16.2.3 Crosspoint switching

If each of N inlets require switching access to any of M outlets, the inlets can be represented as horizontal lines and the outlets as vertical lines, forming a matrix as shown in *Figure 16.4 (a)*. If a switch is introduced at each intersection (crosspoint), connection between any inlet and outlet may be obtained by operating the crosspoint switch at the intersection of the two lines. Systems employing this principle are referred to as crosspoint switching systems.

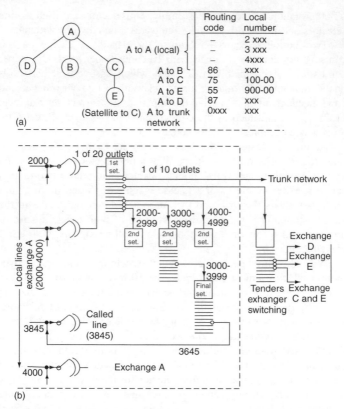

	Routing code	Local number
A to A (local)	–	2 xxx
	–	3 xxx
	–	4xxx
A to B	86	xxx
A to C	75	100-00
A to E	55	900-00
A to D	87	xxx
A to trunk network	0xxx	xxx

(a)

(Satellite to C)

(b)

Figure 16.3 Call routing with Strowger system (non-director); (a) numbering; (b) routing

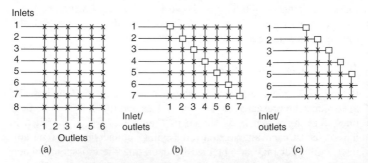

Figure 16.4 Crosspoint switch matrixes: (a) general matrix; (b) square closed matrix; (c) folded matrix. X, switched connections; □, permanent connections

The switches can be electromechanical (cross-bar or relays) or solid state. The cross-bar system uses magnetically deflected horizontal and vertical bars which actuate locking contacts at the crosspoint where they intersect; the bars then release, leaving them ready to set up another call, and the contacts remain locked until released at the conclusion of the call. Relay-switched systems normally employ reed relays, and solid-state systems, bipolar or diode switching. In its simplest form, the matrix provides switching between a two-wire input and a two-wire output, requiring two-pole switching plus any additional poles required for locking or operating logic purposes. Four-wire operation requires four-wire input and output for each channel, and four-pole switching. With two-wire switching the switches must be bidirectional, a condition satisfied by cross-bar or relay switching, but if a solid-state switch is used it also must be bidirectional, favouring a diode gate type of switch and ruling out a number of transistor circuits.

Some degree of electronic control is required with all crosspoint systems to operate the appropriate switches and supervise the maintenance and clearing of the call, and crosspoint switching systems are categorized under the general term of 'electronic switching systems'.

The matrix shown in *Figure 16.4(a)* has completely independent input and output circuits, and requires $N \times M$ crosspoint switches. Frequently, however, each output is connected back to its identically numbered input. This creates a closed system whereby any number can communicate with any other number in the system, and applies to switching between extensions on a PABX, or local calls on an exchange. The number of input and output lines becomes equal (N), and the matrix becomes a square as shown in *Figure 16.4(b)*. The crossconnection of like-numbered inputs and outputs reduces the number of switched crosspoints from N^2 to $N(N-1)$. A call from line 3 to line 6 would operate crosspoint switch 'horizontal 6 vertical 3' and follow the route shown, but could equally well be established by operating switch 'horizontal 3 vertical 6'. The alternative path can be used to provide the return path on a four-wire duplex connection or it can be deleted and the number of switches reduced. Deletion of the alternative path results in the folded matrix shown in *Figure 16.4(c)*, where the number of crosspoint switches is reduced to

$$\frac{N}{2}(N-1)$$

A practical system is likely to require a mixture of closed-system lines (for local numbers) and independent outlets (for calls routed to other sites). The economy in crosspoint switches is only applicable to the closed-system lines.

It is apparent that the number of crosspoint switches increases approximately as N^2, which places a limit on the maximum economic limit of N; an 8000 line exchange would require an 8000×8000 matrix, which with the folded structure would require 32 million switches plus additional switches for outside junctions. The number of switches can be considerably reduced by adopting a three-stage switching network, as shown in *Figure 16.5*. In the arrangement shown the number of outlets and inlets is equal (N). As with the single-stage network the outlets can be connected back to their similarly numbered inlets to form a closed system (as required for calls between terminals connected to the same exchange), or used as separate outlets to route calls to outside exchanges. In a practical application a combination of both is usually required. It should also be

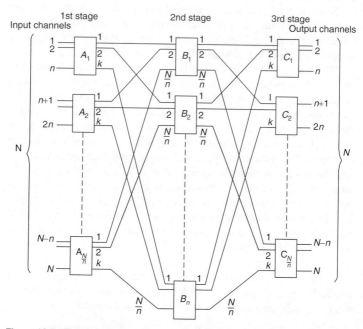

Figure 16.5 Three-stage crosspoint switching: first stage: number of input channels $= N$, number of inputs per A switch $= n$, number of outlets per A switch $= K$, number of A switches $= N/n$; second stage: number of B switch matrices $= K$; third stage: number of output channels $= N$, number of inputs per C switch $= K$, number of outlets per C switch $= n$, number of C switches $= N/n$

appreciated that a number of incoming lines from distant exchanges need to be introduced on the input side, and as the number is likely to be similar to the number of outgoing lines, the number of input and output ports will still be equal.

The number of crosspoints required for the first and third stage are equal (NK each), and for the second stage $K(N/n)^2$. Hence the total number required is

$$2NK + K\left(\frac{N}{n}\right)^2$$

With the single matrix a path can be established between input and output between any two lines provided they are not already engaged. Such a network is said to be 'non-blocking'. This does not apply to the three-stage network, where the number of second-stage matrix switches (K) may be insufficient to guarantee a path for all calls during a peak period. There is therefore a finite possibility of blocking in terms of traffic density, and of the values of N, n and K for a given three-stage switch. To reduce this possibility to zero (i.e. non-blocking) the value of K must be not less than $2n - 1$, which leads to an uneconomic design. The design of an optimum switch configuration for a given number, N, of inlets and outlets, with an anticipated traffic density, therefore involves choosing the values of n and K that provide an acceptable blocking probability with the minimum number of crosspoint switches. Suitable values of n and K for a range of values of N, calculated for a blocking probability of 1 in 500 with a traffic density of 0.1 erlangs per inlet, are given in *Table 16.1*. Note: 0.1 erlangs means that on average the circuit is busy for 10% of the time.

For comparison purposes, the number of crosspoints required with a non-blocking three-stage array and a single-stage switch are also given.

The table shows clearly the economy in switches per line achieved by the three-stage network over the single-stage, and by the blocking network over the non-blocking. This is particularly marked for high values of N, but even with the blocking three-stage array the number of switches per line increases as N increases. The most cost-effective applications of crosspoint switching therefore occur when N is small, that is for small exchanges and PABXs.

High-frequency crosspoint switching is used in satellites. However, the rapid advance of digital switching has overtaken crosspoint switching in the exchange replacement programme.

Table 16.1 Crosspoint switch requirements

N	Three-stage blocking probability (1 in 500 with 0.1 erlangs per terminal)				Three-stage non-blocking		Single stage	
	n	K	Crosspoints	Switches/N	Crosspoints	Switches/N	Crosspoints	Switches/N
32	8	6	480	15	896	28	992	31
128	8	5	2560	20	7680	60	16 256	127
512	16	7	14 336	28	63 488	124	261 362	511
2048	32	10	81 920	40	516 096	252	4.19×10^6	2047
8192	64	15	491 520	60	4.16×10^6	508	67.1×10^6	8191
32 768	128	24	3.18×10^6	96	33.4×10^6	1020	1074×10^6	32 767

193

16.2.4 Digital switching

Time digital switching is achieved by switching a signal from the time slot it occupies in the incoming bit stream to a different time slot in the outgoing bit stream. This process is shown in *Figure 16.6*, applied to a standard primary multiplex 32 channel (E1) 2.048 Mbit/s bit stream. The channel slots are numbered 1 to 32 in the incoming and outgoing streams. Each slot carries an octet (8 bits), and 32 slots (256 bits) form a frame with a duration of 125 μs. The diagram shows the switching of two of the 32 channels from channel 5 to channel 21, and from channel 12 to channel 9. The series of octets appearing in channel 5 are therefore transferred to channel 21, and from channel 12 to channel 9. The following points should be noted:

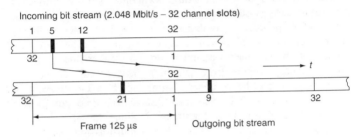

Figure 16.6 Digital time switching

(a) The switching of time slots must involve bit storage and transmission delay, since the bits in each incoming time slot must be stored and await the arrival of the required time slot in the outgoing stream. The average delay can be seen to be half the frame duration, that is 62.5 μs for the 2.048 Mbit/s stream illustrated.

(b) The switched path is unidirectional; duplex operation therefore requires the setting up of a similar path in the opposite direction. For the return path in the case illustrated, time slot 21 in the incoming stream has to be switched to time slot 5 in the outgoing stream.

(c) With duplex operation the time delay of go and return paths can differ, due both to different switching delay and the use of different transmission routing. This is unlikely to cause trouble with digitized speech, but may with data transmission.

(d) Digital switching can be carried out on the standard multiplex bit streams, which are then utilized for both transmission and switching. This simplifies the system architecture and enables channel

194

switching to be achieved without demultiplexing to a baseband signal.

Digital switching as shown in *Figure 16.6* implies the use of a single line into the switch and a single line out of the switch, each carrying a multiplex bit stream of similar format but with the channels displaced in time. This type of switching is therefore referred to as 'time switching', in contrast to the switching described in Sections 16.2.2 and 16.2.3, where the switched outputs appear on physically separate lines, and are referred to as 'space switching'. However, if the output of a digital switch is demultiplexed, the output is split into separate outputs, which may be channels or lower-order multiplex signals, and the digital switching is effectively converted to space switching.

The processes of multiplexing, switching and demultiplexing may all be carried out within the digital switching system, which can therefore produce a combination of time and space switching. Alternatively the digital switching system can transfer the contents of a time slot on one line to a similar slot on another line, producing space switching without time switching. This process is shown in *Figure 16.7*, where the contents of time slot 5 in bit stream 1 are switched to time slot 5 in bit stream 3, and from time slot 13 in bit stream N to time slot 13 in bit stream 2. Since the time slot is not changed, this process does not introduce a transmission delay.

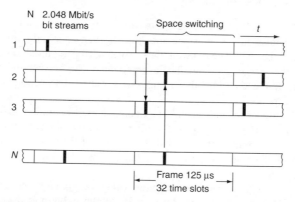

Figure 16.7 Digital space switching

The manner in which time digital switching can be achieved is shown in *Figure 16.8(a)*, and space digital switching in *Figure 16.8(b)*. *Figure 16.8(a)* shows 128 channel 8.192 Mbit/s incoming and outgoing bit streams. To obtain the maximum required

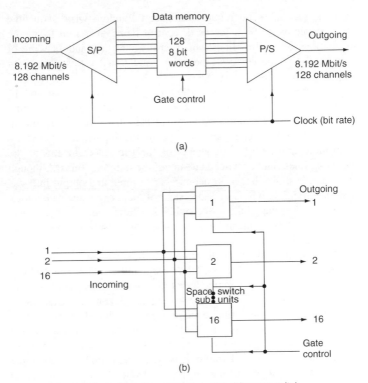

Figure 16.8 Switch configurations: (a) time switch; (b) space switch

switching delay the store must have the capacity to store one complete frame, in this case 128 channels requiring 128 octets of storage. The incoming and outgoing lines could terminate directly on the store, in which case writing and reading will be at bit rate. Although this is satisfactory at lower-level multiplexing (e.g. 2.048 Mbit/s), it is useful at higher rates to convert to parallel operation, with each bit of the octet occupying a separate line and thus reducing the write and read speeds by a factor of eight. The example shown in *Figure 16.8(a)* provides this conversion with series-parallel and parallel-series convertors at the input and output of the store respectively. The store is emptied and refreshed once per frame. Writing can be sequential, in which case each word must be read on to the outgoing bit stream, timed for the time slot to which it is being switched. Alternatively the words can be entered in the order in which they are to appear in the outgoing bit stream, and read out sequentially. These alternatives are referred to as 'sequential write random read' and 'random write sequential read' respectively. Both are used, control being exercised by gating instructions on control lines which ensure that the random write or read results in the correct time delays and transfer to the required outgoing time slots.

The space switch shown in *Figure 16.8(b)* has 16 input and 16 output lines. All input lines are connected to each of 16 switching subunits, and the 16 single outputs of these units form the 16 output lines. Each subunit can gate any one of the inputs, and hence by controlling the gating the contents of the time slots in any of the incoming lines can be switched to any of the outgoing lines. Although the example shown has an equal number of input and output lines (as normally required with a closed system of local exchange switching), this is not essential. If the space switch has fewer outputs than inputs it can be used as a concentrator, and if the outputs exceed the inputs, as an expander.

A single time switch will require multiplex input and output bit streams of $64 \times N$ kbit/s to switch N baseband channels of 64 kbit/s, and single-stage time switching therefore becomes progressively more difficult as N increases. The use of series-parallel conversion as described above assists the actual switching process, but does not reduce the required storage capacity in the switch, or the bit rate of the incoming and outgoing bit streams. Advantage can also be taken of the maximum anticipated traffic density, as measured in erlangs per channel, by employing a concentrator in front of the switch and an expander after the switch. The ratio of concentration must be based on the probability of overload with a given number of traffic units. A ratio of 4:1 would enable 128 channels to be handled on a 32 channel 2.048 Mbit/s bit stream, reducing the switched bit rate by a factor of four.

Further increase in the number of channels to provide the switching capacity required for a large exchange can be obtained by adopting a multiple array, as was shown to be necessary in Section 16.2.3 for crosspoint switching. By combining space and time digital switching, the switching capacity of a single space switch (*Figure 16.8(b)*) can be increased by a factor N, by using N multiplexed inputs and outputs with space switching between them. Two configurations of a three-stage array are possible as shown in *Figure 16.9(a)* and *(b)*: a time switch between two space switches (STS) and a space switch between two time switches (TST). In each case one channel displacement is shown, between slot 2 in bit stream 1 to slot 8 in bit stream N. The same overall switch is achieved by both configurations, the increased

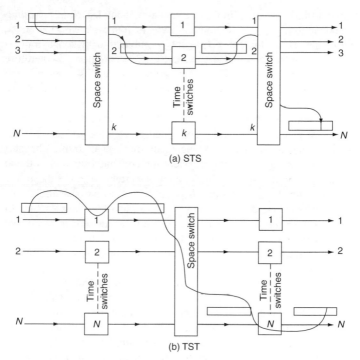

Figure 16.9 Three-stage digital switch: (a) STS; (b) TST

capacity being obtained without increase in the bit rate in any of the switching stages.

Both configurations are used. The STS arrangement enables a small installation requiring only a single-stage time switch to be extended by converting to an STS. It will also be seen that the space switches in an STS array can have different numbers of inlets and outlets, and can therefore fulfil the roles of concentrator and expander as well as providing the space switch. The STS array has therefore found favour in commercial systems for smaller exchanges. For larger systems the TST configuration has been found more cost effective. A further factor is the positioning of the concentrator. In earlier systems the tendency was to operate two-wire analogue up to the switch, the analogue-to-digital conversion, two- to four-wire operation, and concentration preceding switching. The increasing proportion of direct baseband digital transmission, and the decreasing cost of digital processing, have made it advantageous to move these processes out towards the terminal, with the eventual aim, using ISDN, of providing analogue-to-digital conversion and two- to four-wire working at the terminal, and multiplexing (possibly combined with concentration) between the terminal and the exchange. Concentration has therefore occurred before the digital switch is reached, and the fact that the TST array cannot combine concentration and space switching is not a disadvantage.

For transmission purposes the E1 2.048 Mbit/s primary multiplex uses channel 0 for stabilization and synchronization and channel 16 for separate channel signalling, reducing the number of traffic channels to 30. In the digital switching system the contents of channels 0 and 16 are removed and used in the control system to guide the traffic channels through the switch. It is possible therefore to use all 32 time slots for switching data. This, however, also involves the switching of the signalling bits in channel 16 to ensure that they remain associated with their own traffic channel. The switch also has to establish the 'return' path which can be done either simultaneously with the initial 'go' path, or independently.

Most of the major telecommunications manufacturers produce and market their own digital switching equipment. This differs considerably in detail design, but it all conforms to the general switching principles explained above. All the equipment tends to use LSI extensively for multiplexing, demultiplexing, switching and analogue-to-digital conversion. It differs in the detailed methods of concentration and expansion, switching format, the means of dealing with overload, and the steps taken to deal with security. Security in this context is the protection of the system against the breakdown of a piece of critical common equipment, such as a processor which could affect the operation of a high percentage of the exchange.

16.2.5 Digital switching control

Both space and time digital switching is achieved by applying correctly timed gating impulses to the various stages of the switch array. The gating signals are derived from the control circuitry which is in turn software driven from a central processing unit. Detail design of the control system differs considerably between the various manufacturers, and the software may represent several years of concentrated development and a significant proportion of the investment in the system. Some details will be found in the references already quoted, and this section is limited to a general summary of the functions of the control system.

It is apparent from the description of the switching process that no direct relation exists between the dialled number and the numerical sequence required to gate the switch arrays. The control system therefore has to translate the incoming address information into appropriate gating signals to ensure that the outgoing signals are in the correct slot and bit stream. If the call is to a distant exchange the control circuits must translate the incoming routing information to a numerical sequence representing the optimum route for the call. In a digital exchange, information on the required routing to all exchanges can be stored in the memory of the SPC (Stored Program Control). The SPC also enables alternative routing to be available, which becomes operative if the first-choice route has no free channels.

The signals coming into the exchange may be analogue or digital. If the signal is analogue, the address information will normally be loop interrupt (10 ips) or two-frequency MF4. This has to be processed for analogue-to-digital conversion, and the address signalling translated to a binary signal.

In a three-element array the control must allocate a free path through the centre element, and must be aware of the status of every time slot under its control. It should also be appreciated that the control system has to establish a return path for duplex operation, and also to ensure that routing information, separated at the switch input, is reintroduced as required at the output. For a local call the output multiplex line (1 to N) together with the time slot define the called local number. A stage of expansion may be required, which will be the inverse of the concentration carried out at the input. A call to another exchange will require at least the called subscriber's number to be passed on, and where the trunk network is involved, routing digits as well. The control system therefore has to ensure that this information is inserted in the correct time slots of the correct multiplex output line.

Other functions of the control system are call clearing and supervision, and call charging. The adoption of SPC enables these functions to be carried out with greater sophistication. With adequate storage, details of every call and its cost can be recorded and made available, with billing, to the subscriber. Flexible call supervision enables call diversion, call barring, calling number advice, and other useful features to be added as they become available, usually by additions to the software. Systems differ in the manner in which the control is assembled. Some systems concentrate it in one central control, others prefer to distribute it to provide separate processors for various parts of the system, with the processors communicating with one another.

16.2.6 Traffic units and Erlang's formula

The routing of information in a telecommunications system is based on the concept of a number of inputs, all having an intermittent demand, which have to be distributed between a smaller number of outlets. If each input had its own output channel there would theoretically be no risk of a lost call, but such a system would be uneconomic. As the ratio of outlets to inlets decreases, the risk of a caller finding all outlets engaged increases, and the grade of service deteriorates. Owing to the random nature of the demand imposed by the inputs, the estimation of grade of service is a problem in probability.

The demand can be represented in terms of 'traffic units', or 'erlangs', the number of traffic units being the total of the average demands of the inputs, that is

$$A = \frac{Ct}{T} \text{ erlangs}$$

where C is the number of calls in time T and t is the average duration of calls. For the purpose of estimating grade of service the peak-period demand must be used to obtain the value of A.

The economic design of a system must ensure that a number of outlets or switches N carries the maximum possible number of traffic units, that is that the ratio A/N is as high as possible.

The permissible ratio of A/N must be based on an acceptable grade of service and this can be defined in terms of the probability of all the outlets being in use, resulting in a lost call. A grade of service of B is then defined as providing a probability of B in 1 of a call being lost at peak periods. It is apparent that the higher the value of N, the lower will be the value of B, due to the distribution of the load between a greater number of outlets. This was investigated using probability theory by Erlang, resulting in Erlang's loss formula:

$$\text{Grade of service} = \frac{A^N/N!}{1 + A + A^2/2! + A^3/3! + A^4/4! + \ldots + A/N!}$$

where A is the number of traffic units (erlangs) and N the number of outlets.

The formula assumes that traffic arises in a random manner. The manner in which A/N varies with N over a range of grades of service, as predicted by Erlang's formula, is shown in *Figure 16.10*. From these curves it can be seen that for a high grade of service, say 0.005,

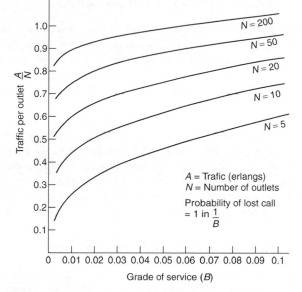

Figure 16.10 Grade of service: A = traffic (erlangs); N = number of outlets; probability of lost call = 1 in $1/B$

representing a 1 in 200 probability of a lost call, the traffic that can be carried per outlet is increased over four-fold for an increase in N from 5 to 20.

The curves refer to a full availability group, that is the total traffic has access to all outlets. An advantage derived from electronic switching is the ability to scan and select an unused outlet from a large number of outlets, that is to use a high value of N. This adaptability was not obtained from the older mechanical switching systems, which are limited to 10 outlets for two-motion switches and 25 or 50 for uniselectors. The efficiency difference, in terms of A/N between 10 and 200 outlets for a grade of service of 0.005, is about 1 to 2. To overcome this loss of efficiency a system known as grading has to be used, whereby the later-choice outlets are made available to a larger number of inlets.

A disadvantage of using large groups (high values of N) is the increased vulnerability to overload. This can also be seen from the curves of *Figure 16.10*. A system designed for a grade of service of 0.005 would degrade to about 0.024 for a 10% overload with 200 outlets, but to only about 0.007 with five outlets.

16.3 Private branch exchanges (PBXs)

The process of central interconnection was first satisfied by a manual private branch exchange (PBX) where an operator made the necessary cross connections by means of plugs and sockets, or keys. Although the routing of all incoming calls via the operator has certain advantages, the need for operator intervention for internal calls, or for dialled-out calls, is inconvenient and time wasting.

The modern PBX allows internal calls and calls via the PSTN to be dialled and the connections made automatically by a central switching unit. Most PBXs, however, retain the advantage of operator intervention of incoming calls by providing an operator console, to which incoming calls can be routed, enabling them to be connected to an appropriate terminal. The duty of operator can then conveniently be combined with that of secretary or receptionist.

16.3.1 Analogue private branch exchanges (PBXs)

PBXs provide automatic centralized switching for the two-wire inputs from the terminals and connections to the PSN. They extend from very small installations with fewer than 10 terminals, to installations catering for several thousand terminals.

PBX technology has followed developments in automatic switching as used in the public switched network, starting with Strowger electromagnetic switches, followed by crosspoint techniques, with some of the more recent equipment employing digital switching.

In a public exchange the number of lines is usually so high that several tandem stages of switching are required, whereas with a PBX the numbers are very much smaller. The switching process is simplified when the numbers are sufficiently small to require only one stage of switching.

In a PBX using crosspoint switching, each outlet will normally be connected to its equivalent input, so that any terminal can call any other terminal. If the number of terminals is N, the number of crosspoint connections for this purpose is $N(N - 1)$ for a square matrix, which can be halved if a folded matrix is used (Section 16.2.3). The number of crosspoint switches therefore increases rapidly as N increases, setting a limit to the number of terminals that can be handled with a single-matrix system. The single matrix has the advantage of simplicity and freedom from blocking, and provides an elegant solution for the smaller PBX.

To reduce the number of crosspoint switches in a larger PBX a three-stage switching array is necessary (Section 16.2.3). The reduction in crosspoint switches is achieved at the expense of complexity and a possibility of blocking, the actual reduction depending on the probability of blocking accepted for a given traffic density.

The actual crosspoint switching can be achieved by using an inter-locking cross-bar with leaf contacts, relays, reed switches or solid-state switches. All are suitable for handling analogue transmissions. The first three provide contact switching and could therefore handle digital signals, but the other circuitry in the PBX may contain elements such as transmission bridges which cannot handle digital signals. Also, with an analogue system, the inputs from terminals and exchange will be two wire, and the switch will provide transmission switching for two wires only.

16.3.2 Digital PBXs

The analogue switching systems are adequate for business organizations whose requirements are limited to speech communication and facilities such as telex, facsimile, etc. adapted by the use of modems, to transmission on the speech-frequency PSTN. Digital PBXs directly interface with digital exchanges and ISDN, and facilitate networking between associated PBXs.

For a digital switched PBX each terminal has its own time slot, and connection of two terminals involves the switching of time slots to enable each terminal to transmit and receive in its own time slot. The way in which this is achieved can be seen by considering typical equipment, in conjunction with the general description of digital switching given in Section 16.2.4. The PBX has an associated operator console that has an alphanumeric display which provides the operator with information on the status of calls being handled, and a keyboard. Incoming calls are normally received by the operator, who determines the extension required by the caller and completes the setting up of the call. To do this, and to enable direct through dialling to the extension where this facility is available, the incoming call is routed via one of the input ports, and is processed by the same digital switching sequence as applies to calls originating from the terminals.

16.3.3 Direct dialling in (DDI)

Direct dialling in (DDI) allows callers to ring into a PBX, but includes digits that direct the call to a specific extension. In the USA, this is known as direct inward dialling (DID). Generally DDI is used in conjunction with multi-line ISDN.

The advantage of DDI to the caller is that they can dial direct to the person or department they wish to speak to. The advantage to the called person is that they can set up a conference call, divert the call or transfer all calls to another extension. The PBX retains all its usual features as far as internal calls are concerned.

16.4 Next generation PBXs (NG-PBX)

First generation PBXs are now mostly obsolete, but were based on analogue switching. Second generation PBXs are the current type and are based on digital switching. Some second generation PBXs provide computer telephony integration (CTI) through a CTI link that connects the PBX and the LAN server. These PBXs with CTI are sometimes called hybrids and are intended to extend the life of second generation PBXs.

The third generation of PBX, called next generation PBXs, are computer-based. A computer telephony (CT) server houses cards with PBX functionality and this is connected to the LAN. Some companies believe next generation PBX is an IP-PBX, which is an IP based PBX that uses IP telephony. Voice packets would then be switched by the IP network, rather than by a separate switch.

The general view of the industry is that CT servers are third generation PBXs and IP-PBXs are fourth generation. However, some companies claim to be providing fourth generation PBXs now. Nortel produce an IP telephone (i2004) and a PBX called Enterprise Edge that has LAN and PSTN connections (I have been told that this uses Norstar digital telephones too). Lucent's 'Argent Office', known in the UK as BT Fusion PBX, is reported to have similar functionality.

PBX related standards are produced by ECMA have produced QSIG which is the replacement for DPNSS, some PBXs have this interface already. ECTF have produced H.100 and H.110 as bus standards within CT servers. EITF have produced SIP which is supposed to be simpler than the ITU-T H.323. Then of course there are ETSI and ANSI.

16.4.1 CT server-based PBX

An CT server-based PBX, sometimes known as a personal computer PBX (PC-PBX), is a server fitted with cards for ISDN interfacing, extension interfacing, fax modem, automatic speech recognition, etc.

The cards draw power and receive commands via the PC's bus, but they are connected by a separate bus (ribbon cable) to provide a synchronous path for the digitized speech. There are a number of synchronous bus 'standards', including Compact PCI, SC-bus, VME, etc., but H.100 and H.110 (ECTF standards) are fast becoming popular. The H.100 bus uses a 68-wire ribbon cable to carry 32, 2.048 Mbit/s lines; this gives up to 4096 time slots.

A basic PBX costs less than CT servers. However, CT servers have many features considered as 'add-ons' in the PBX market. If a PBX with high functionality is required, a CT server works out cheaper. PBXs are capable of about 70 functions. Those least supported by a CT server are voice mail, call recording, fax gateway and operator services.

16.4.2 Internet protocol PBX (IP-PBX)

In the future, Voice over IP will grow and provide PBX functionality, the switching being part of the IP network rather than a separate physical item. However, some leading edge users have found that using IP telephony over the same LAN as the data causes problems. During high data usage, the speech quality falls below commercially acceptable levels.

Internet protocol (IP) is a packet-based method of transmitting and routing data over a common circuit. PBX systems that are linked to CTI servers enable call related data to be displayed. If the PBX was itself a server, the CTI functionality could be contained within one box. Transmitting speech and data in a common form is attractive because only one link needs to be established in order to exchange data whilst having a conversation. Next Generation PBXs could be IP-based and not circuit switched; basically PBX software rather than hardware would provide the PBX functionality. Soft PBXs (IP-PBXs) are expected to develop over the 2001–2004 timescale. Packet switching using IP will be used in future networks and these may use version 6 (IPv6) because it allows more addresses and more control of the packet routing.

In an IP switched network, Gateways are used to connect the telephone network to a LAN or to the Internet and provide address translation, alarm functions and management. Gatekeepers are software functions used to provide call control, call routing and other basic PBX services such as call transfer. A DHCP server is also required to allocate individual IP addresses. These can all be provided by a single server (example: Phoneware SBX Server, running 'SBX mixer' for conferencing and 'SBX gate' for interfacing).

Speech carried over an IP network suffers from packet loss, delay and jitter. Over an Intranet, which is controlled, packet loss is less than 0.1%. Over the Internet, packet loss can be as high as 4%. The effect of this on speech depends on how it is encoded; G.711 is more resilient to packet loss than G.723.1, which has high compression. Jitter on an Intranet (an internal, controlled network) is about 50 ms, but can be as high as 200 ms on the Internet, which is almost unacceptable. Delays of 200 ms or more give poor results.

A virtual private network is a network over a physically shared medium that behaves like a dedicated circuit. This can be an ATM circuit or a GSM circuit that carries packets for a number of users but the calls are separated at the ends of the network by having different address headers. An IP-VPN is an IP-based VPN. VPNs can be used instead of private circuits to lower the cost of long distance routes.

16.4.3 Call centres

In call centres, groups of operators work together for selling or advising on products. The hardware is an automatic call distributor (ACD) which is a PBX that answers calls and places callers in a queue. As an operator becomes free, the longest waiting call is connected through. Call centre functionality is often provided by CTI. The operator is able to input data to a computer terminal during the call and this is associated with the call details (such as calling line identity).

Some call centres are used for making outgoing calls; these have lists of numbers that the computer polls, ringing one number after another and then connecting an operator when a called person answers.

16.4.4 Wireless PBX

The wireless office means different things to different people. To some people it means using mobile phones (such as GSM or DECT) to replace the desk telephone. To other people it means replacing the LAN cable with HiperLan/2 (5 GHz) and using Bluetooth (at 2.4 GHz) to link GSM phones to the laptop PC. Some people think it means both. The increasing use of the mobile phone has affected PBX use. Many PBXs now have a DECT interface or allow calls to be diverted to a mobile phone.

16.4.5 Centrex

Centrex uses the local public exchange to carry out call switching. Each extension is individually wired to the exchange, which uses local network resources of the telecommunications service provider but frees the user from PBX purchase and accommodation issues. Internal calls within the Centrex group are free and can be set up using short code dialling so that the user is unaware that his calls are passing outside the company premises.

Mobile Centrex (currently used in Denmark) performs the same function, except that the GSM network carries out the call switching. Call tariffs are adjusted so that 'internal' calls are nil or low cost. The network cost is recovered from 'external' calls. Short code dialling is available for 'internal' calls. Because of the number of potential users in a small area, it has only been possible to set up the service in small companies (<200 users).

17 Switched networks

The telephone network is made flexible and efficient by the use of switching. A switching hierarchy begins with the local network. Telephone exchanges are connected in a tiered fashion, so that many local exchanges feed into fewer tandem exchanges, which are cross connected and feed into fewer higher level exchanges, and so on. The proportion of international calls is small, so these are connected at the highest level. The whole hierarchical design is based on the probability of a call. The higher the probability, the lower the tier through which the call is routed. Physically, these exchanges could all be adjacent or in the same building (in a major city).

17.1 Local networks

At the lowest level of the switching hierarchy is the local exchange. Every customer has a dedicated, unswitched circuit feeding into the exchange. A circuit is most likely to be a copper pair, used for basic telephony (i.e. telephone, facsimile and modem data). A few copper pairs may carry ISDN, HDSL or ADSL circuits. Larger customers may have optical fibre cable installed.

It is not possible for each customer to have a separate cable; a small feeder cable at the customer's premises is combined with others at distribution points. The distribution cable may have a number of distribution points along its length. It is at these points where further customers' feed cables are joined to pairs in the distribution cable.

Primary connection points (PCP), sometimes called cabinets, are provided at convenient points. These allow pairs from several distribution cables to be jointed to pairs in a much larger cable which goes directly to the exchange (see *Figure 17.1*). The cost (per circuit) in terms of cable manufacture, installation and duct space is reduced by using the largest cable available to suit the number of customers in a particular area. Spare pairs are allowed for, since a telecommunications company cannot afford to replace a cable if just a few pairs become faulty.

The local exchange feeds traffic to the tandem exchange hierarchy. If the trunk network is not provided by the same company as the local network, as is becoming the case in a number of countries including the USA and the UK, the local exchange performs the first part of routing the call to the appropriate trunk carrier. A local exchange will

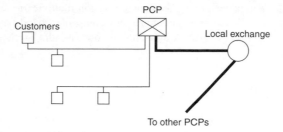

Figure 17.1 Local network cabling

be sited close to the centre of an urban district, with cables from customers feeding in from all sides. Similarly, a tandem exchange will be sited in the centre of a group of local exchanges. The aim is to keep the cable requirements to a minimum (although it is the ducts, installation and maintenance that are expensive, not the cable itself). The general idea is illustrated in *Figure 17.2*.

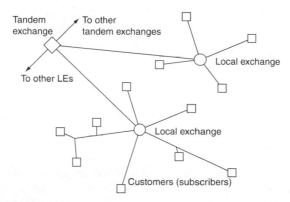

Figure 17.2 Location of exchange sites

In the USA, Regional Bell companies feed traffic to AT&T, MCI, Sprint and others. Usually in the USA local calls are not charged; the toll exchange is owned by the trunk carrier so charges are only applied to calls entering their network. This may change because the increased use of the Internet, which are local calls and not charged, has resulted in severe congestion.

In the UK, BT provides both local and trunk networks, but other trunk network providers can use the BT local network. Local calls are deemed to cover a small group of exchanges and these are charged, although at a much lower rate. Local networks are also provided by cable TV companies, using frequency division multiplexing on their coaxial feeds. Some telecommunications companies avoid having to access the BT local network by using microwave radio for their local access.

17.2 National networks

It is the national networks that contain the tandem exchange hierarchy. Again, the physical location is designed to keep the overall cabling requirements to a minimum. An exchange hierarchy diagram is given in *Figure 17.3*.

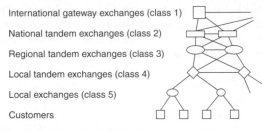

International gateway exchanges (class 1)

National tandem exchanges (class 2)

Regional tandem exchanges (class 3)

Local tandem exchanges (class 4)

Local exchanges (class 5)

Customers

Figure 17.3 Exchange hierarchy diagram

Links from the local exchange (class 5 office) to the local tandem exchange (class 4 office) is usually a digital link having a minimum data rate of 1.544 Mbit/s (2.048 Mbit/s in Europe). Generally, links to the tandem exchanges at the top of the hierarchy have very high data rates, transmitted over long distances using optical fibre links.

Each long distance carrier company will have its own hierarchy of switched networks. There are no links between the networks; only links to the local exchanges at either end of a network. Thus, a call will pass through the local exchange and then into (say) an MCI network. At the other end of the country, the call will come from the MCI network and pass through a second local exchange before reaching the end customer. Each carrier will have international links routed through exchanges at the top of the hierarchy (class 1 office).

17.2.1 The intelligent network

The intelligent network (IN) is a signalling network that uses the SS7 protocol (see Chapter 19) in order to create additional services. The SS7 protocol and the intelligent network are used for:

- basic call set-up, management, and tear down;
- wireless services such as personal communications services (PCS), wireless roaming and mobile subscriber authentication;
- local number portability (LNP) and carrier pre-select (see Section 17.2.2, below);
- call routing and billing for toll-free 0800 (800/888 in the USA) and premium rate 0945 (900 in the USA) fixed-line services;
- enhanced call features such as call forwarding, selective call barring, calling party name/number display (calling line identity) and three-way calling;
- efficient and secure world-wide telecommunications.

Call signalling data is transmitted in packets over a dedicated network between local and tandem exchanges, via a packet switching hub. A database containing, for example, details of the call routing for toll-free calls is also connected to the signalling network.

17.2.2 Carrier pre-select

Carrier pre-select allows customers to use alternative service providers, for local, long-distance or international calls, or all three. It is usually used for international calls and implemented by dialling a prefix number before the required telephone number, but in some cases it may be implemented automatically by prior arrangement. Sometimes a unit that plugs in between the telephone and the wall socket is used (this automatically dials the prefix number when a call is initiated). Carrier pre-select relies on the Intelligent Network because the prefix number must be translated into call routing to another service provider.

17.2.3 Internet call routing

The telephone network was designed to handle voice calls that had a certain call pattern, usually voice calls are of short duration and the number of calls peaks mid-morning and mid-afternoon. The mid-morning peak is the highest and is known as busy-hour.

Calls to internet service providers (ISPs) do not follow the voice traffic calling pattern. The calls to ISPs are long duration, usually several hours, and traffic volumes show less pronounced peaks. Switching ISP traffic through the PSTN is inefficient because the data itself arrives in bursts, yet a modem must remain connected and send idle signals between the bursts. More importantly, ISP traffic can cause blocking; this is where the switch reaches full capacity and voice calls cannot be completed. To overcome these difficulties, calls to ISPs are diverted at the local switch to a packet switching network, which overlays the PSTN. A modem in the exchange terminates the call and forwards data over an Internet Protocol (IP) packet network. The PSTN to IP network signalling interface is a computer server, known as a gateway.

17.3 International networks

International links are provided by normal cable if a suitable route exists (e.g. between Germany and France). Islands, such as Great Britain or Ireland, have to use submarine cable or satellite links.

17.3.1 Submarine cable links

Submarine cables are a critical element of international telecommunications networks. Modern fibre optic cable systems have the capability to transmit vast amounts of data at a low cost relative to copper cable and satellites. Both incumbent and emerging carriers have recognized the flexibility of submarine cables in satisfying their rapidly growing infrastructure requirements.

The increase in demand for submarine cable capacity is being driven primarily by data traffic from Internet use. Technologies such as wave division multiplexing (WDM), synchronous digital hierarchy (SDH) and optical fibre amplifiers will allow virtually limitless transmission of information across the globe.

Originally, copper-based coaxial cables were used for submarine links, using frequency division multiplexing (FDM) technology to provide a large number of circuits. These were superseded by submarine optical fibre cable, first proposed by AT&T in 1979. This, and other cables like it, used digital multiplexing technology to provide many more circuits than previously available using FDM systems. The TAT-8 cable went into service in December 1988, and connected the USA, the UK and France; it had a maximum transmission capacity of 560 Mbit/s. Other 560 Mbit/s cables placed into service at about the same time were: TAT-9, TAT-10 and TAT-11, TPC-4, APC, China-Japan and Columbus-2.

Advancements in SDH technology allowed submarine systems to break the 560 Mbit/s barrier to achieve up to 10 Gbit/s on a single fibre. CANTAT-3 was one of the first SDH cables operating at 5 Gbit/s carrying 32 STM-1 circuits. CANTAT-3 connects Canada, Denmark, Germany and the UK, using 89 repeaters on a 44 600 fibre-km system.

Other early SDH cables included TAT-12 and TAT-13, TPC-5, APCN, FLAG, Se-Me-We-3 and Americas-1.

WDM was introduced and allowed the capacity of existing systems to be increased without having to replace the physical cable. The first submarine cable systems that employed WDM technology were TAT-12/13 (two- to four-channel upgrade), Gemini (six- to 12-channel upgrade), AC-1 (eight- to 16-channel upgrade) and Columbus-3 (four-channel upgrade).

SDH ring technology allows the automatic restoration of circuits and reduces the need for long-distance carriers to have diverse cable and satellite systems, in case of failure. Submarine cable systems are now constructed with the cable in a ring configuration. In the event of a fibre break, the direction of traffic reverses around the ring. The first major submarine ring-system was the TAT-12/13 combination. Some 50% of the capacity on TAT-12/13 is reserved for restoration purposes. The TAT-12 cable provides the northern part of the ring and TAT-13 provides the southern portion.

A critical technological advance has been the development of systems that do not require repeaters. These cables are laid in relatively short spans (1000 kilometres or less) without any electricity feed to repeaters. Electricity feeds running alongside the cables are a fault liability. The planned Caribbean ARCOS-1 system will use repeaterless technology on 22 of 24 links between North, South and Central America.

17.3.2 Plesiochronous operation and slip

The concept of plesiochronous working is a technique employed to enable digital streams of approximately the same rate to be multiplexed together to form a single higher data rate stream, and involves the process of justification whereby identifiable dummy bits are added to a tributary stream to make it truly synchronous.

A quite separate problem occurs, for instance with the 64 kbit/s channel time slots in a primary stream, if a received signal is processed by sampling with a non-synchronous clock when quite clearly bit errors and pulse degradation could occur. The loss of information when the signal rate is faster than the sampling clock, or the repeated information when the signal rate is the slower, is called SLIP. A single slip of 8 bits from a 64 kbit/s stream, representing one sample in a PCM voice channel, is called an OCTET SLIP. A slip of one frame of a 2048 kbit/s digital stream, representing simultaneous slips of one sample (octet) of 30 PCM channels is called a FRAME SLIP.

In the BT UK network a technique of mutual synchronization has been adopted to control slip. A triplicated caesium reference clock with a stability of $1 : 10^{11}$ is the top level of a four-tier synchronization network. Control between levels of the hierarchy is one way, that is from a higher to a lower level, and is thus effective only at the lower level. In any one level, control between nodes is bilateral, whilst at the lowest level (e.g. digital local exchange) master-slave synchronization phase-locks the exchange clock to the bit rate of a nominated synchronization link. The exchange clock is likely to be an oven-controlled quartz oscillator, within a phase-locked loop, and having a stability around $1 : 10^{10}$ per day. The loop filter will have a long time constant of 1000 s or so.

Control information to achieve synchronization is contained in time slot TS0 of nominated 2048 kbit/s links; these signals speed up or retard an exchange master-slave clock, and in practice the exchange scans all its incoming synchronization links and acts according to a majority decision.

Clearly the same problem of slip can occur in international digital communications be they via cable or satellite. ITU-T Rec. 811 specifies the timing associated with nodes terminating international links, including the timing associated with TDMA (Time Division Multiple Access) satellites. Recommendation G822 concerns performance targets for controlled slip rate on an international digital link. In it, transmission without significant impairment on a 64 kbit/s connection is achieved with no more than five slips per 24 hours; a slip rate of more than five per 24 hours but less than 30 per hour is considered acceptable for speech but not for other services, whilst more than 30 slips per hour is unacceptable for any service.

17.3.3 Transmission standards

The public switched telephone networks world-wide are designed to meet standards laid down by ITU-T and although ITU-T recommendations strictly apply only to international links, in practice national networks operate to the same design parameters. This ensures that the points of origin and receipt of an international call do not affect its transmission quality.

In the digital transmission system, performance impairment is no longer primarily related to the attenuation/frequency characteristic of the transmission medium since digital regenerators reconstitute the data stream to its original form, ideally at each repeater station, and the final receiving equipment needs only to detect the presence or absence

215

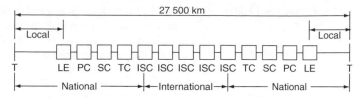

Figure 17.4 T, reference point (ITU-T Rec. 1411); ISC, international switching centre; TC, tertiary centre; PC, primary centre; LE, local exchange

of pulses. Impairments in a digital system are basically random by nature; noise, cross-talk effects, equipment imperfections, impulsive electrical interference are typical sources which can cause bit errors. The performance of a digital system is therefore assessed in relation to a hypothetical reference connection (HRX) between two reference points. ITU-T G821 uses the model HRX shown in *Figure 17.4*; other shorter HRXs are defined in G104 and draft G81X. G821, first published in 1980 and since revised, gives the following objectives for an international HRDX at 64 kbit/s:

Performance classification   *Objective*

(a) Degraded minutes   Less than 10% of 1 minute intervals to
                         have a BER worse than 1×10^{-6}
(b) Severely errored   Less than 0.2% of 1 second intervals to
    seconds   have a BER worse than 1×10^{-3}
(c) Errored seconds   Less than 8% of 1 second intervals to
                         have any errors, that is 92% error-free
                         seconds

Three different quality classifications, based on practical digital transmission systems, have been identified as local, medium and high grades, and these generally relate to location within the network. Local grade covers systems operating into local exchanges over a mix to links between LEs and into the bearer system, while high grade embraces systems over long-haul bearer circuits and international links, and which may be via wire, fibre optic, microwave line-of-sight radio or satellite. The allocation of error objectives between the various grades of the HRX is as follows:

Local grade (two ends)   15% of degraded minutes and
                                errored seconds objectives to
                                each end

| Medium grade (two ends) | 15% of degraded minutes and errored seconds objectives to each end |
| High grade (25 000 km link) | 40% allocation (equivalent to 0.0016% per kilometre) |

Various mathematical models predict the error performance of an ISDN. The problem of relating direct measurements of BER at higher multiplex bit rates (e.g. 140 Mbit/s) to the performance at the 64 kbit/s base rate must be considered.

Section 17.3.2 introduced the concept of slip performance, control of which is clearly related to agreed international time standard clocks. The USA, Italy and Australia have a single (triplicated) caesium standard like the UK; other networks, however, may have a number of synchronized regions each with an independent reference clock. Although national network topologies differ, methods of slip control are similar and overall international synchronization is maintained within the recommendations of ITU-T G822. Subdivided into the same grades as above, these are:

Local grade	<5 slips/24 hours	for 98.9% of time
Medium grade	>5 slips/24 hours ⎱ <30 slips/hour ⎰	for less than 1% of time
High grade	>30 slips/hour	for less than 0.1% of time

The figures, as before, relate to a 64 kbit/s data rate.

18 Repeaters

Repeater is a misleading term but is sometimes used to describe both analogue amplifiers and digital regenerators; both are used to compensate for cable losses. The output of a 'repeater' should be a copy of the signal at the input at the far end of the cable. This is where analogue and digital systems differ. An analogue amplifier simply increases the amplitude of a signal at its input – this includes any noise and crosstalk induced into the cable prior to the amplifier. A regenerator not only restores the digital signal to its maximum level, it also retimes the signal to remove any timing jitter at the output caused by noise signals at the input.

18.1 Analogue amplifiers

The unavoidable frequency-dependent attenuation incurred in copper twisted pair and coaxial cables used for analogue transmissions is compensated by the insertion of amplifiers – or repeaters, as they have come to be called – at intervals along the cable. These repeaters must be designed for low noise, low intermodulation distortion and freedom from overload under peak signal conditions. Integrated circuit amplifiers are currently available for this application, although the circuit may use discrete transistors.

A long wideband link will require a large number of repeaters and failure of one repeater could result in the loss of thousands of channels. The application of quality control is of paramount importance throughout design and manufacture. Various additional safeguards such as use of redundant components or even modules automatically brought into service if failure should occur are commonly employed.

A typical noise target, per repeater, is about 4 pWOp (4 pW psophometric) made up of 2.7 pW thermal and 1.3 pW intermodulation noise achieved with an amplifier noise factor (NF) of 14 dB, but in practice will be less than this.

Repeaters used in coaxial systems carrying broadcast TV programmes (i.e. studio/transmitter links) will include phase equalizers as well as attenuation equalizers in order to maintain the required video-frequency transient response.

The cable losses are temperature sensitive (approximately 0.2% per °C) and over a 2 km section of the above cable this could amount to an unacceptable 1 dB per 10°C at 12.5 MHz. The usual method of compensation is to monitor the level of a pilot signal (12.435 kHz) which can be isolated via a crystal filter, and then amplified, detected and compared with a stable reference DC. The difference signal is amplified to bias a thermistor whose resistance determines the loss in the negative feedback network of the repeater. Other pilot tones may be used for supervisory purposes along a chain of repeaters.

A stringent requirement between the amplifiers of a bidirectional repeater employed in a two-wire system, is that of cross-talk, particularly for channels used for broadcast programme material (e.g. leased to the BBC). A typical target is 170 dB attenuation between the input of one amplifier and the output of the other, and this is only achieved by good design of screening and strict attention to elimination of common earth currents.

An amplifier is a unidirectional device, and it is impractical simply to parallel two amplifiers in a two-wire or coaxial circuit. One solution is to use hybrid transformers at each repeater as in *Figure 18.1.*

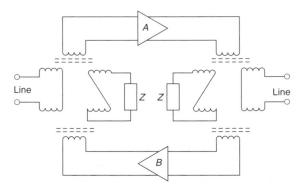

Figure 18.1 $Z =$ line matching impedance

Where a four-wire system is used, then the above arrangement is only necessary at terminal stations, since each pair of the four-wire system can have its own train of unidirectional amplifiers.

Repeaters are powered by a constant DC current fed along the inner of a coaxial cable; typically the current is 40–50 mA, the repeater operates at about 15 volts, and the supply voltage is 100 volts. Isolation filters are necessary to prevent signal currents bypassing a repeater via the supply and these must obviously have 'in-band' attenuation in excess of repeater gain. One arrangement is shown in *Figure 18.2.*

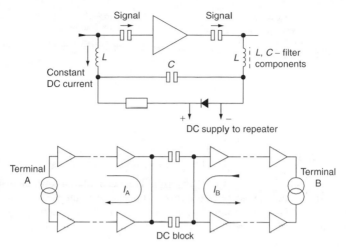

Figure 18.2

The lower diagram of *Figure 18.2* shows how power is 'turned round' to return to power sources A and B. In this all filters have been omitted for clarity; the signal coupling between 'A' and 'B' sections could alternatively be by transformer coupling. Repeater amplifiers have of necessity to be well protected against voltage surges along the line due to lightning strikes and other forms of interference. Zener diodes, non-linear resistors and gas tubes are all in common use for this purpose.

18.2 Digital regenerators

The digital pulses of a PCM system suffer distortion when transmitted over a length of cable due to the combined effects of attenuation and group delay versus frequency. Providing data has not deteriorated to the extent that the presence or absence of a pulse cannot reliably be detected, it can be reconstituted without error to its original amplitude and shape.

The equipment which does this is the digital regenerator which is inserted at regular intervals along a data cable. A simplified block diagram of a regenerator used with HDB3 signals is shown in *Figure 18.3*.

The input data stream is phase equalized and amplified, and then applied to a clock pulse generator and phase split to separate +ve and −ve pulse generators, the outputs of which are recombined to line. A typical 140 Mbit/s regenerator could have an input threshold of

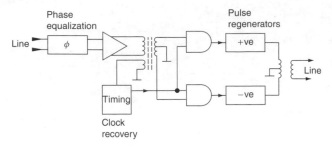

Figure 18.3

−50 dBm and an output of +23 dBm. This technique allows almost distortion-free and noise-free transmission, since white noise is not cumulative along the route. In this respect it differs totally from analogue transmission in which the signal/noise ratio deteriorates with length of route.

Digital regenerators are constant-current powered via isolation filters just as in the repeaters of Section 18.1, the current typically being 50 mA at a voltage drop per regenerator of 15–20 volts.

Although the pulses detected by a regenerator are reconstituted to their original form, if any incoming pulses are undetected then clearly the resulting gaps in the outgoing data stream constitute errors. The measurement of bit error ratio (BER) over a stated period of time is an important function of PCM system testing.

18.2.1 Submerged digital regenerators

Power feed is by constant current (typically 500 mA) and the power sources, which must obviously be land based, may have voltages up to 10 kV. Extreme precautions are taken to ensure adequate protection of all components, for example gas tubes at input and output, breakdown diodes across the amplifier low-voltage supply, individual diodes in transistor emitter circuits to protect base/emitter junctions. Clearly all components are conservatively rated to guarantee a long working life.

18.3 Optical repeaters

Pulses of light carried over optical fibre can be amplified, without conversion to electrical signals and back, by using Erbium doped fibre amplifiers. These work in a manner similar to that of a laser. They comprise a 50 m length of Erbium doped fibre that is coiled to keep its size as small as possible and a light source that is coupled into the fibre. These are shown in *Figure 18.4*. The purpose of the light source is to provide the energy for the amplifier, typically between 1 and 100 mW.

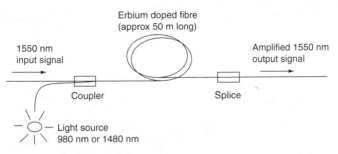

Figure 18.4 Erbium doped fibre amplifier

The optical amplifier works by using the light source to raise the energy of electrons within the Erbium doped fibre. The presence of the Erbium creates an energy state that is metastable (semi-stable). Any electrons in the metastable state are triggered into returning to the ground state energy by the presence of a photon, provided that its energy is equal to that of the metastable state. For each electron returning to the ground state, a photon with equal energy to the metastable state is generated. Thus one photon into the optical amplifier can generate many more, as excited electrons release their energy in the form of further photons, as illustrated in *Figure 18.5*.

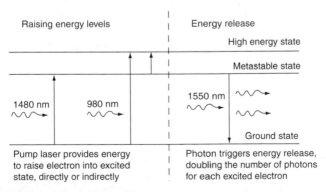

Figure 18.5 The photon-multiplying effect

In Erbium, the metastable state has an energy equal to a photon with a wavelength of 1550 nm. Thus Erbium doped fibre can be used to make a 1550 nm optical amplifier. The light source for this amplifier must have a wavelength of less than 1550 nm (i.e. each light source photon must have more energy than the metastable state). Typical light sources are at wavelengths of 980 or 1480 nm.

Optical amplifiers using Erbium doped fibre are being used on the latest submarine cables. This choice restricts the optical wavelength to 1550 nm. Work is underway to produce an optical amplifier that works at 1300 nm.

18.4 Radio relay

A radio relay receives signals on one antenna and transmits on a second. If the antennae are highly directional (which is the case for microwave dishes) and pointing in the opposite direction, it is possible to receive and transmit at the same frequency. This could be a simple case of amplifying the signal received on one antenna and outputting it to the second antenna, particularly if the amplifier gain is less than about 20 dB. If the signal is digitally modulated, any noise picked up on the receive antenna will be amplified and retransmitted; this will cause jitter of the demodulated signal's timing. It may be better to demodulate the signal, re-time the recovered data and then use this to modulate a second transmitter.

The transmitter must be operating on a different frequency from the receiver if good isolation between the receive antenna and the transmit antenna is not possible. The received signal could be converted to baseband and then used to modulate another carrier. If the two carrier signals are in different frequency bands, frequency translation circuits can be used; these use a mixer with a local oscillator signal operating at a frequency that is the difference between the receive and transmit frequencies.

19 Signalling

Signalling describes the data that must be sent with a message to ensure its arrival at the correct destination. The simplest example is perhaps a long distance telephone call: (1) dial tone is signalled to the subscriber, telling him the exchange is ready to receive call routing information; (2) the dial pulses or tones signal the exchange on how to route the call; (3) the ringing or engaged tones let the caller know the status or progress of the call; (4) at the distant end ringing current signals the telephone of the called subscriber, to indicate an incoming call.

In the telephone call example, other telephone exchanges must be given call routing information. Inter-exchange signalling can use the same channel as the call, but only if the signalling information is transmitted before the call is connected through or if the signalling frequencies are outside the speech bandwidth. Where call carrying channels are digitally multiplexed onto a single circuit, a separate data channel is used to carry the signalling information.

The use of a separate data channel has allowed the development of additional services. Call diversion, caller identity, ring-back when free, etc., are just a few.

19.1 Signalling requirements

In the modern network some automatic means of signalling is required to interconnect subscriber 'A' via a series of switching centres to subscriber 'B'; to supervise the call along its route; to give caller 'A' status information such as a dial tone, ringing tone, engaged tone; and to initiate and record call charges.

Signalling data can be transmitted by a variety of techniques which may have to be compatible with one another since, for example, the dialling speed of dial telephones is still 10 pps (pulses per second) even though the call may be routed via high bit rate digital transmission links. The following methods are all in use:

(a) in a two-wire DC loop, the direction and level of current;
(b) pulse duration (DC);
(c) pulse combination (DC);
(d) frequency of an AC signal;

(e) combination of frequencies;
(f) binary or other code.

The application of these methods is described in the following sections.

19.2 Local line signalling

This has developed as essentially a two-wire DC technique with the exception of ringing current and standard tones. Originally, in manual exchanges, these tones were produced electromechanically and were

Ringing tone	133 Hz	interrupted
Engaged tone	400 Hz	interrupted
Out of order	400 Hz	continuous
Ringing current	17 Hz	(at 75 volts)

The signalling sequence initiated by the caller is shown in *Figure 19.1* and is referred to as loop disconnect dialling.

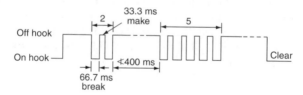

Figure 19.1

The DC current which flows when the caller is 'off hook' is interrupted at 10 pps by the dial pulses (or the equivalent in a modern DTMF telephone). DC signalling is limited distance-wise by resistive voltage drop and cannot be applied directly to carrier systems.

Modern DTMF handsets and electronic exchanges have progressed towards eliminating the relatively high-power requirements of the above system, at the same time giving the customer improved signalling speed resulting in quicker connections. The frequencies used for signalling are illustrated by *Figure 19.2*. In the need of compatibility, dialled digits are stored electronically in the DTMF keypad phone and transmitted to line as a 10 pps impulse loop disconnect when operating into a Strowger exchange.

A form of line signalling not apparent to the subscriber is known as CLASS signalling. This uses V.21 signalling (the same as a 300 baud modem). CLASS signalling is used to signal the caller identity

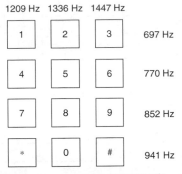

Two tones are transmitted concurrently
for each key press. Pressing a 7 causes
1209 Hz and 852 Hz tones to be transmitted.

Figure 19.2 DTMF Frequencies

to the subscriber. The technique used in the UK is to send the caller
identity before any ringing is applied, so that the number is available
upon the first ring. At present, the called subscriber is alerted to the
presence of a CLASS signal by a line polarity reversal. In the USA,
the caller identity is sent between the first and second rings.

Further applications for CLASS signalling include non-telephone
call applications such as reading electricity and gas meters. In these
applications, a line reversal would not be applied to the line. Instead
the receiving modem will monitor the line for tones and must receive
a matching 'address' for it to respond.

19.3 Channel associated signalling (CAS)

All of the systems described in this section, whereby signalling information related to a traffic circuit is carried along the same transmission
channel (physical path) as the traffic circuit, are referred to as channel
associated signalling.

In analogue systems signalling transmitters/receivers are provided
on a per circuit basis; in the case of line signalling (e.g. SF) they
are provided as a permanent connection and for interregister signalling (e.g. MF) switched in as required. *Figure 19.3* shows this
in schematic form.

An out-of-band signal at 3825 Hz is used exclusively in carrier
systems for supervisory purposes and offers the advantage of continuous supervision during a call. ITU-T Rec. Q351 refers to this system
as R-2 signalling with the facilities listed in *Table 19.1*.

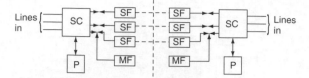

Figure 19.3 SC, switch centre; P, processor; SF, single-frequency signalling; MF, multifrequency signalling

Table 19.1

Circuit condition	Direction	
	Forward	Return
Idle	Tone on	Tone on
Seized	off	on
Answered	off	off
Clear back	off	on
Released	on	on or off
Blocked	on	off

19.3.1 Signalling System No. 4

ITU-T SS4 (in the UK SSAC4) is an example of an end-to-end signalling system using 2VF at 2040 Hz and 2400 Hz. It is used for line and interregister signalling, the latter with serial transmission in binary code.

If we call 2040 Hz 'x' which is equivalent to binary '0' and 2400 Hz 'y' equivalent to binary '1', a four-element code provides 16 characters (i.e. 10 digits plus six supervisory) as listed in *Table 19.2*.

In this system each line signal is made up of an initial prefix 'P' followed by a control element. The prefix consists of both frequencies, and the control one frequency with tone durations as follows:

$$P = 150 \pm 30 \text{ ms}$$

$$x \text{ and } y \text{ each } 100 \pm 20 \text{ ms}$$

Further supervisory signals are in the form of XX or YY which are of 350 ± 70 ms duration. Thus we have, for example, in the forward direction

Terminal seizing	PX
Transit seizing	PY
Digits	as *Table 19.2*
Clear forward	PXX
Forward transfer	PYY

Table 19.2

	1	2	3	4
1	y	y	y	x
2	y	y	x	y
3	y	y	x	x
4	y	x	y	y
5	y	x	y	x
6	y	x	x	y
7	y	x	x	x
8	x	y	y	y
9	x	y	y	x
0	x	y	x	y
Call operator code 11	x	y	x	x
Call operator code 12	x	x	y	y
Spare code	x	x	y	x
Incoming half echo suppression reqd	x	x	x	y
End of pulsing	x	x	x	x
Spare	y	y	y	y

and in the backward direction

Proceed to send	X
International transit	Y
Engaged	PX
Answer	PY
Acknowledge	P

19.3.2 Signalling System No. 5

ITU-T SS5 (UK SSAC10/SSMF1) describes a system using 2VF (2400 Hz and 2600 Hz) for line signalling on a link-by-link basis plus a 2MF using two-out-of-six frequency code at 9 digit/s for inter-register signalling. The six frequencies are spaced 200 Hz apart from 700 Hz to 1700 Hz. In the USA a similar, but not identical, signalling system is known as R-1.

The ITU-T SS5 code is

1	$700 + 900$ Hz
2	$700 + 1100$
3	$900 + 1100$
4	$700 + 1300$
5	$900 + 1300$
6	$1100 + 1300$
7	$700 + 1500$
8	$900 + 1500$

9	$1100 + 1500$
0	$1300 + 1500$

Supervisory codes are

Prefix digit sequence	$1100 + 1700$ Hz
End of digit sequence	$1500 + 1700$
Operator code 11	$700 + 1700$
Operator code 12	$900 + 1700$
	$700 + 1100$
Payphone coin control	$1100 + 1700$
	$700 + 1700$

ITU-T Rec. Q361 defines a two-out-of-six MF signalling system using two different sets of six frequencies for forward and return directions. These are spaced at 120 Hz from 540 Hz to 1980 Hz and give enhanced, though similar, facilities to those provided by ITU-T SS5.

In all signalling systems involving tones, some form of filtering is clearly necessary to separate out the various frequencies. Originally in the earlier systems this would have been done with passive *LC* filters, but in more recent designs other techniques avoiding the use of inductors are employed. Obviously the power level of in-band signals has to be kept within specified limits, and these are defined in the relevant ITU-T recommendations.

19.4 Common-channel signalling (CCS)

In common-channel signalling the signalling information is separated from groups of voice trunks and sent in serial binary form over a single path dedicated only to signalling. In FDM systems this data is interfaced via modems to a 2400 bit/s signalling link. In the 30 channel PCM system the bits allocated to TS16 are extracted from the frame and sent on a separate channel. Several hundred voice channels may have all their signalling information combined in one common signalling channel. *Figure 19.4* is a schematic arrangement.

The common-channel signalling information must contain address data identifying the voice channel to which it relates, and CCS is only possible between exchanges using processor control so that the correct association between signal and relevant voice channel can be re-established at the receiving exchange. The physical route taken by the CCS channel can differ from that of the traffic circuits so that the signalling network may be economically optimized. The common-signal

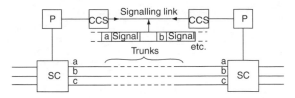

Figure 19.4 SC, switch centre; P, processor; CCS, common-channel signalling (unit)

channel can also be advantageously used to carry non-traffic-related signals required for operational/maintenance purposes.

19.4.1 ITU-T Signalling System No. 7 (Q701-707, Q721-725, Q741)

SS No. 7 is designed for digital communications networks, including ISDN, which link stored program control exchanges, and although optimized for 64 kbit/s bearers it is intended to be usable at lower rates over analogue bearers (e.g. 4.8 kbit/s). SS No. 7 is known as C7 (based on the old title of CCITT Signalling System No. 7) and has a structure based on the ISO seven-layer model.

The system is constructed on a modular basis and consists of a four-level hierarchy in two main parts: the message transfer part (MTP) covers levels 1–3 and the application-dependent user part (i.e. telephone, data terminal, ISDN) is level 4. *Figure 19.5* is a schematic of the overall structure, in which the four levels correspond closely to the first four levels of the seven-layer OSI model.

Level 1 defines the physical and functional characteristics of the signalling data link (normally 64 kbit/s) and the access to it. Normally this is TS16 in the 30 channel PCM system.

Level 2 carries out the signalling link function including adaptation from the processor to the 64 kbit/s data stream. Bit fields for error detection and correction and flags for signal unit (SU) delimitation are added to this level.

Level 3 carries out the signal network functions, that is directs a message to the correct signal link, or user part, reconfigures routes after a failure, and sends data regarding abnormalities in the signal network.

Level 4 is the user part, for example TUP (Telephone User Part), DUP (Data User Part (modem)).

ITU-T Q701 defines the SIGNAL MESSAGE as an assembly of information determined at level 3 or level 4 pertaining to a call

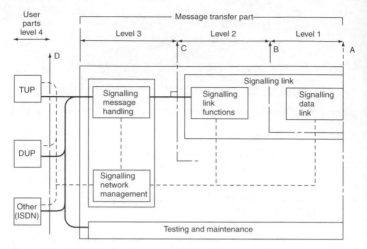

Figure 19.5 TUP, telephone user part; DUP, data user part; -•-•-•-•, signalling message flow; - - - -, controls and indications

management function which is transferred as one entity by the message transfer function. Each message contains service information which identifies the source user part etc. The SIGNALLING INFORMATION contains data on call control signals, type and format of the message, an address label and supervisory information. This allows level 3 to route the message to its correct destination.

The signalling data link (level 1) is a full duplex link which can operate over terrestrial or satellite channels. Apart from TS16 in the 2048 kbit/s stream, TS67–70 in the higher-order 8 Mbit/s multiplex can also be used, the signalling bit rate being 64 kbit/s in both cases.

The signalling terminal (level 2) has a number of functions to fulfil and these are:

1 *SU delimitation and alignment*:
   Error detection (using cyclic redundancy code $x^{16} + x^{12} + x^5 + 1$)
   Error correction (ARQ-see Section 18.10)
2 *First-time alignment*:
   Signalling link monitor (of bit error rate)
   Signal message flow control (e.g. prevention of queuing or blocking).

The basic format of a message signal unit (MSU) is shown in *Figure 19.6* in which:

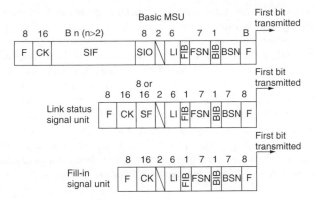

Figure 19.6

F   = Flag (01111110)
CK   = Check bits
SIF = Signalling information field (this is variable, as generated by the user part)
SIO = Service information octet
LI   = Length indicator
FIB = Forward indicator bit
BIB = Backward indicator bit
BSN = Backward sequence number
FSN = Forward sequence number (this is achieved by incrementing the previous FSN by 1 in modulo 128)

A positive acknowledgement (ACK) is indicated by the receive terminal assigning the received FSN value to the next BSN sent in the backward direction. The BSN then remains unchanged until a further MSU is acknowledged. A negative acknowledgement (NACK) merely inverts the BIB.

The S10 is divided into a service indicator and a subservice field; the service indicator is 0100 for telephone, 0101 for ISDN, whilst the 4 bits of the subservice field are used to discriminate between national and international networks.

The SIF has an integral number of octets of user data at present from 2 to 62, but the eventual maximum as per ITU-T recommendations is 272.

Two other types of signal unit are used when no messages are available for transmission and as a result SIF is not required. These are a fill-in signal unit (FISU) and a link status signal unit (LSSU); the former contains forward and backward indicator bits and sequence numbers whilst the latter contains an additional status field used for signalling link control.

Signalling information for non-telephony services that in due course will be available to PSTN customers has been considered in the form of a protocol referred to as transaction capability (TC). This is intended to be used in a variety of applications including customer-to-customer data transfer, mobile network support and other enhanced and value-added services. ITU-T TC recommendations are in Q771–774 and are fully compatible with SS No. 7 requirements.

Signalling System No. 7, even from this brief summary, can be seen to be a complex signalling system, still evolving as it is expanded to include new customer and network features.

19.5 SS7 and the intelligent network (IN)

Common channel signalling system No. 7 (i.e., SS7 or C7) is a global standard that defines the procedures and protocol by which network elements in the public switched telephone network (PSTN) exchange information over a digital signalling network, to effect mobile (cellular) and fixed-line call set-up, routing and control. The ITU definition of SS7 allows for national variants; for example, those defined by the American National Standards Institute (ANSI) and the European Telecommunications Standards Institute (ETSI).

The SS7 protocol is used by the Intelligent Network (IN, see Chapter 17), which is a signalling network that is used to provide additional services. The SS7 protocol and the intelligent network are used for:

- basic call set-up, management, and tear down;
- wireless services such as personal communications services (PCS), wireless roaming and mobile subscriber authentication;
- local number portability (LNP) and carrier pre-select;
- toll-free 0800 (800/888 in the USA) and premium rate 0945 (900 in the USA) fixed-line services;
- enhanced call features such as call forwarding, selective call barring, calling party name/number display (calling line identity) and three-way calling;
- efficient and secure world-wide telecommunications.

19.5.1 Signalling links

SS7 messages are exchanged between network elements over 64 kbit/s (56 kbit/s in the USA) bi-directional channels, which are called signalling links. Dedicated channels are used to carry the signalling data, rather than in-band over voice channels. Compared to in-band signalling, dedicated signalling channels provide:

- faster call set-up times;
- efficient use of voice circuits;
- support for intelligent network (IN) services, particularly where voice circuits are not required (e.g., database systems);
- improved fraud control.

19.5.2 The SS7 network architecture

Each signalling point in the SS7 network is uniquely identified by a numeric point code. Point codes are carried in signalling messages exchanged between signalling points to identify the source and destination of each message. Each signalling point uses a routing table to select the appropriate signalling path for each message. There are three kinds of signalling points in the SS7 network: SSP (service switching point), STP (signal transfer point) and SCP (service control point). These are shown in the SS7 network architecture diagram, in *Figure 19.7*.

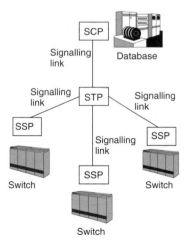

Figure 19.7 SS7 network architecture

An SSP is a switch that originates, terminates, or tandem routes calls. An SSP sends signalling messages to other SSPs to set-up, manage, and release voice circuits. An SSP may also send a query message to a central database (at an SCP) to determine how to route a call (this applies to premium rate and toll-free calls). The SCP replies with the appropriate routing number(s). An alternate routing number may be used by the SSP if the primary number is busy or otherwise unavailable.

An STP is a packet switch that is used to route network traffic between signalling points; the routing information is contained in the SS7 message. The STP acts as a network hub, which removes the need for direct links between signalling points and thereby improves the utilization of the SS7 network. An STP can also act as a 'firewall' gateway for exchanging SS7 messages with other networks.

19.6 Inter-PBX signalling

Early drafts of the ITU-T recommendations for digital subscriber signalling had gaps in the definitions, which led to problems in PBX signalling. Various national operators, including BT (the national telecommunications operator in the UK) produced their own standards. The BT standard was called Digital Access Subscriber System No. 1 (DASS1). Later an inter-PBX standard called Digital Private Network Signalling System (DPNSS) was developed from DASS1. The development of DPNSS highlighted problems with DASS1, which were corrected in a new 'standard': DASS2. The DASS2 standard allows for ISDN access defined by I.430 and I.431.

Development of the ITU-T digital subscriber signalling system No. 1. (DSS 1), under recommendations Q.920–Q.931, has continued and will eventually replace all the national PBX signalling standards.

20 Non-switched networks (LANs and WANs)

Data can be shared between computers if they are connected together by a circuit. Modems allow computers in people's homes and offices to be connected to the Internet, which could be considered a wide area network (WAN). However, this chapter will concentrate on private networks used between offices of one company, or a group of companies. A local area network (LAN) is one where all the computers are on a single site. A WAN is two or more LANs on separate sites, connected together.

20.1 Local-area networks (LANs)

LANs arose primarily to provide data communication between a number of terminals in a single office complex. This followed the move towards diversification of computing and information storage between the terminals to replace the earlier practice of employing one large central computer. All terminals share a common bus or ring system to which access must be gained in order to transmit a message. Packet switching techniques are employed so that each packet of information utilizes the bus for only a very short period of time, circulating to all other terminals, but only communicating with the terminal whose address appears at the beginning of the packet. The system is therefore simplex; if a reply is required it can be provided by the called terminal sending a separate packet to the calling terminal. Also the sharing of a common bus or ring relies on all packets being sufficiently short for no terminal to hog the system, and these two restrictions make the system basically unsuitable for speech transmission.

LAN systems differ in respect of the bus configurations, the means of gaining access, and the avoidance of collision when more than one caller is seeking access at the same time. The alternatives include bus and ring configurations using twisted-pair cables, coaxial cables or fibre optic cables, and the methods of access include carrier sense multiple access (CSMA) (which may include collision detection (CD)), empty slot, and token-passing techniques. A single LAN may operate over distances typically between 0.1 and 10 km, usually on a single site, with means for interworking with the PSN or associated LANs.

236

The basic principles of operation of some typical systems are described in the following sections.

20.1.1 10 Mbit/s thick coaxial Ethernet (10 Base 5)

The bus for Thick Coaxial Ethernet is a passive coaxial cable with a maximum length of 500 m and able to accommodate up to 100 terminal stations fed in along its length. Avoidance of reflections along the cable is important, and the cable is terminated at each end with its characteristic impedance. The terminals must be teed into the cable without creating an impedance irregularity, and this is achieved by using a transceiver to couple the terminal (via up to 50 m of feeder cable) to a high-impedance tee connection to the cable. Additionally the cable is marked at 2.5 m intervals and connection made only at these points, which is claimed to reduce the possibility of unwanted reflections. The arrangement is shown in *Figure 20.1*.

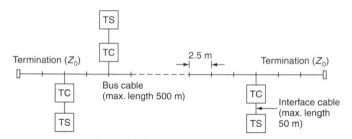

Figure 20.1   Single Ethernet segment: TC, transceiver; TS, terminal station

If the total number of terminals exceeds 100, or the terminals are distributed over an area that cannot be served by a single 500 m bus, additional bus networks can be interlinked, but repeaters are required at the points where the bus systems are linked, as shown in *Figure 20.2*.

20.1.2 10 Mbit/s thin Ethernet (10 Base 2)

Instead of using a thick inflexible coaxial cable, which has to be pierced at precise locations in order to connect workstations, a thin coaxial system was devised. This uses 50 Ω coaxial cable with a braid outer conductor for flexibility. Connections to computers are made using Bayonet Niell and Concelman (BNC) connectors, provided that there is at least 50 cm of cable between each pair, as illustrated in *Figure 20.3*.

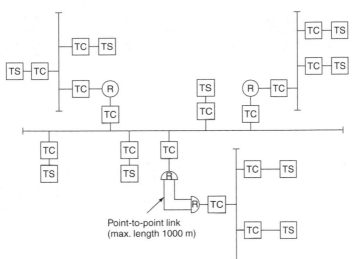

Figure 20.2 Multisegment Ethernet: TS, terminal station; TC, transceiver; R, repeater

The maximum length of a 10 Base 2 Ethernet is 185 m, although repeaters can be used to extend the range to 925 m. Up to 30 workstations can be connected to a single segment and terminating resistors are required at the ends.

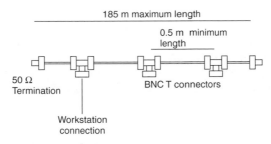

Figure 20.3 Thin coaxial Ethernet

20.1.3 10 Mbit/s twisted pair Ethernet (10 Base T)

Twisted pair cabling is attractive because of its simplicity. Unlike the 10 Base 5 and 10 Base 2 systems, the 10 Base T system uses star topology. Each workstation is connected to a central point called a hub. There are usually several hubs in a system, connected in a tree

238

and branch arrangement. The computer server connects to one hub and this is in turn connected several more. Those hubs may be connected to further hubs or may be connected to workstations, as illustrated in *Figure 20.4*.

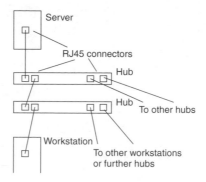

Figure 20.4 Twisted pair Ethernet

The connectors used with this type of Ethernet are RJ45. Two pairs are used although half duplex transmission is used. At the workstation, terminals 1 and 2 are transmit and terminals 3 and 6 are receive. At the hub, terminals 1 and 2 are for receive and terminals 3 and 6 are transmit. Thus connecting a workstation to a hub requires a straight connection. Connecting two hubs together requires a crossover connection.

Up to 1024 computers can be interconnected using twisted pair Ethernet, but there must be no more than 4 hubs connected in series. It is recommended that the number of workstations should be far lower than this; a reasonable limit is 50. Unshielded twisted pair (UTP) category 5 (CAT5) is recommended for all LAN installations because it has the ability to carry 100 Mbit/s data rates (100 base T).

20.1.4 Data over Ethernet

Access to the bus employs CSMA, and the terminal wishing to use the bus must first check that it is unoccupied. Having seized the bus, the send terminal transmits a packet having the format shown in *Figure 20.5*, preceded by a 64 bit preamble to establish stabilization and synchronization. The first two groups of octets provide information on the destination and source addresses, each requiring six octets (48 bits). The long address is necessary because a unique physical address is allocated to each terminal by the system manufacturer, so

Preamble	Destination	Source	Type	Data	Frame check sequence
8 octets	8 octets	8 octets	2 octets	48–1500 octets	4 octets

Figure 20.5 Ethernet packet format

that in theory Ethernet terminals may communicate on an international scale. In addition a simpler 'multicast' address can be allocated which can be used for intercommunication between a group of related stations.

Since seizure requires that the bus appears to be unoccupied, collision can only occur due to the time interval that exists between a transmission starting from one terminal and reaching and being recognized by other terminals (a delay of the order of 4 μs per kilometre). Should another terminal gain access during this short period, a collision results. During transmission an Ethernet terminal monitors the line, and should it receive a signal differing from its transmitted signal, it assumes a collision and stops transmitting. An algorithm is then employed to control reaccess from the contending stations. This is referred to as collision detection (CD).

Ethernet normally operates at 10 Mbit/s and the longest packet (1526 octets) therefore only occupies the network for about 1.25 ms.

A number of simplified systems operating on the Ethernet principle are available. Economy has been achieved by operating at reduced bit rates, sometime using cheaper twisted-pair cable, and in some instances by dispensing with the CD facility [5].

Ethernet is now covered by an international standard ISO 8802/3.

20.1.5 100 Mbit/s Ethernet systems

A twisted pair Ethernet operating at 100 Mbit/s (100 Base T) is almost identical to 10 Base T, except that it requires special hubs. Many of these hubs also support 10 Base T and this allows parts of the LAN to operate at different speeds. A variant is called 100 Base TX which allows full duplex operation.

Optical fibre Ethernet (100 Base FX) uses two strands of 62.5 micron core fibre and allows full duplex working. The large core allows cheap plastic push-in connectors to be used for the workstations.

20.1.6 Hubs, repeaters, bridges and routers

Hubs are a building block of twisted pair and optical fibre LANs. Repeaters, bridges, and routers are hardware designed to link one

LAN to another. Bridges, routers and repeaters all perform the same basic function, but for different applications.

Hubs were briefly described in Section 20.1.3. They provide a distribution node for twisted pair and optical fibre LANs. Both twisted pair and optical fibre systems use point-to-point transmission. The hub allows several (typically nine) circuits to exchange data. Like the repeater, all data is re-transmitted, with no address checking. Data received on any one circuit will be output on all the other circuits.

Repeaters are the simplest LAN connection device; they are used to connect two identical type of LANs. Their purpose is to extend the length of a LAN and they do this by regenerating *all* signals received from one LAN for output on the other. They are bi-directional and are available to carry electrical or optical signals.

Bridges are intelligent repeaters and provide some degree of filtering. A bridge will build up an address table by monitoring the traffic and deciding whether a certain address is on one side or the other. A data packet received from one LAN but addressed to a workstation on the other LAN will be transmitted. Conversely, if the data packet is received on one LAN and is addressed to a workstation on the same LAN, it will not be transmitted.

The most basic bridges require the same type of LAN on either side, but more sophisticated bridges are available to provide interconnection between dissimilar types. The MAC sublayer address is used to determine the workstation's location, but this is different for each type of LAN, i.e. CSMA/CD addressing differs from token ring addressing. A translating bridge can interpret addresses on different types of LAN and allows a CSMA/CD bus to connect to a token ring.

A remote bridge is designed to connect LANs via a wide area network. These bridges operate in pairs, one on each LAN with an intervening WAN. Data on one LAN is converted into X.25, Frame Relay, ATM, or whatever WAN protocol is required.

Routers are more sophisticated than bridges and can build up knowledge of the whole network by interacting with other routers. Routers are like bridges in that they build up address tables and only pass traffic from one LAN to another if the workstation being addressed is on the second LAN. Unlike bridges, the address that the router uses is at the LAN protocol level. One of the most popular LAN protocols is TCP/IP, although Xerox's XNS, Digital's DECnet, AppleTalk and others are available. Thus, for example, a TCP/IP address will be used rather than a MAC sublayer address. Most routers require the same protocol (e.g. TCP/IP) for both LANs, but some provide a translation facility. Routers are often used to provide WAN interconnectivity.

20.1.7 Cambridge or slotted ring

The Cambridge Ring was developed by the Cambridge University Computing Laboratories as a means of interconnecting a wide range of computers and similar devices at data rates in excess of those available at the time using conventional telecommunications techniques. The Cambridge Ring LAN uses the circulating empty-slot principle, with mini-packets circulating round the ring with a bit rate of 10 Mbit/s.

The ring configuration is shown in *Figure 20.6*. Transmission is unidirectional around the ring, which employs telephone cable with repeaters spaced at intervals of 100 to 300 m. Terminal stations access the ring at the repeater points, and access boxes provide an interface between host computers and the terminal, enabling the computers to send or receive data over the ring.

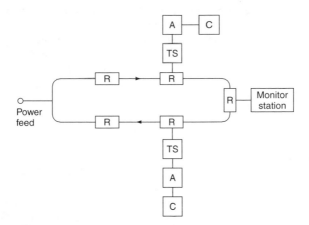

Figure 20.6 Cambridge Ring: R, repeater; TS, terminal station; A, access box; C, host computer

Each station can be programmed to accept incoming packets from any source, selected sources only, or no packets, by means of its source select register (SSR).

The packet format is shown in *Figure 20.7*, and this is first set up and transmitted from the monitor station, which then monitors the information in the packet each time it circulates past.

242

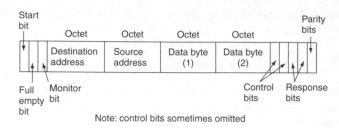

Figure 20.7

Note: control bits sometimes omitted

An empty packet shows '0' in the full/empty bit slot. As the packet passes a station attempting to gain access, the station can seize the packet by changing this bit to '1', indicating that the packet is full. It also sets the monitor bit to 1, inserts the destination and source address (each of 8 bits), followed by the data (two octets), and sets the response bits to 11. Operation around the ring is then as follows:

(a) The destination address is checked by each terminal as the packet passes. If the address differs from the terminal address the packet continues until it finds the correct destination.

(b) The addressed terminal next checks the source address and compares this with its SSR. If this is programmed to receive no calls, the response bits are changed to 10. If programmed to receive from a different transmitting station the response bits are changed to 00. If the terminal is inoperative and ignores the call, the response bits remain unchanged at 11. In all of these cases the packet continues round the ring, the content of the response bits indicating to the monitor and transmit stations the fate of the call.

(c) If the addressed station accepts the call, it changes the response bits to 01 after accepting the data.

(d) When the packet passes the monitor station, the monitor bit is changed to 0. A packet arriving at the monitor station with the monitor bit at '0' will have circulated twice and the monitor station will cancel the packet and initiate a new empty packet.

(e) When the packet arrives at the transmit terminal, the state of the response bits will indicate whether the packet has been accepted, or the reason for rejection. If accepted, the data is removed, the monitor response and full/empty bits restored, and the packet reinstated in the ring as an empty packet. If the packet was rejected the transmit terminal can decide whether to try again or send to a different terminal.

The requirements of the slotted ring LAN are covered by the international specification ISO 8802/7.

20.1.8 Token-passing techniques

Token passing is an alternative means of obtaining access without collision, and can be applied to either bus or ring LANs. The token consists of a unique bit sequence which must not occur in any data stream.

Access can only be obtained by seizing the token. A station wishing to transmit must therefore check and retransmit the token with a delay of 1 bit. Only when the complete sequence has been checked can the token be verified as available and seized. The delay allows the last bit to be changed, indicating that the token has been seized. Seizure of the token must be followed by transmission of destination and source addresses, and the data. After completion of the data message, the transmitting station restores the token which passes to the following stations.

When operated on a ring, transmission is normally baseband and circulates round the ring until the destination address is reached. If the addressed station is able to accept the packet, it deletes the packet information, thus indicating to the transmit terminal that the data has been delivered. If, on arrival back at the transmit station, the data is still in the packet, the call has been unsuccessful and the caller can make a further attempt after a delay. In either case the packet is emptied and the token reinstated.

The token-passing ring is covered by specification ISO 8802/5, which deals with a baseband system operating at 4 Mbit/s. Higher rates are available but are not included in the international standard.

Operation as a bus system is more complex, as the token has to circulate to all stations, that is on both sides of the sending station. This means that the bus system must be bidirectional, involving a transponder at the terminations to reverse the direction of transmission. This can be achieved either by using separate cables for the two directions or by using a broadband carrier system with different carrier frequencies for the two directions. The token-passing bus is dealt with in specification ISO 8802/4.

With all token-passing bus systems, once the token has been seized the amount of data transmitted is not limited by a packet format. Systems do, however, normally impose a limit to avoid hogging by one terminal.

Token-passing LANs have been developed and utilized mainly for manufacturing process control.

244

20.1.9 Fibre distributed data interchange (FDDI)

FDDI is typically used on a university campus or a business premises and gives a100 Mbps wide backbone for LANs. Up to 500 nodes can be supported, spaced no more than 2 km apart. Two fibre optic rings are used and these are arranged so that data is counter rotating – the same data travels on both fibre rings, one clockwise and the other anti-clockwise. This topology gives some fault tolerance – the dual ring is converted to single ring if a fibre fails, see Figure 20.8. The maximum ring length is 100 km as a dual ring (200 km as a single ring).

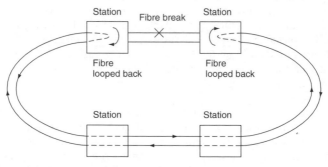

Figure 20.8 Dual ring fault tolerance

ANSI is the main standards authority for FDDI and they have given it the designation X3T12. The International Standards Organization have also 'standardized' it under the designation ISO 9314. The FDDI standard was developed from the token ring standard, IEEE802.5. In the ANSI standard there are four key components: MAC – Media Access Control; PHY – physical layer; PMD – physical media dependent; and SMT – station management protocol. The MAC component defines addressing, scheduling, data routing and communication using protocols such as TCP/IP. The PHY component handles encoding/decoding, NRZI modulation and clock synchronization. The PMD component handles analogue base-band transmission between nodes – fibre and copper. The SMT component handles ring management including neighbour identification, fault detection and reconfiguration.

FDDI cables usually employ multi-mode fibre with a 62.5 micron core and 125 micron cladding diameter. The 62.5/125 fibre is favoured because low-cost LED/photodiode technology can be used to drive/detect light in wider fibre. However, 50/125, 85/125 and single mode fibre can also be supported. Four fibres are used in each cable (2 transmit and 2 receive) although the installation of spares is recommended for replacement of faulty fibres.

It is also possible to use FDDI formatted data over copper, which has the acronym CuDDI. This uses copper unshielded twisted pair (UTP) or shielded twisted pair (STP) cables for connecting to local terminals. The copper cable can be up to 100 metres in length and ANSI standard TP-PMD (twisted pair – physical medium dependent) applies. The advantages of using twisted pair cable are that it is low cost and that installation and termination are simpler. Also, copper-based transceivers are cheaper, smaller and require less power compared to fibre-based systems.

Unlike token rings defined by IEEE 802.5, there is no active ring monitor. Instead, each ring interface has its own clock synchronized to incoming data. The outgoing data is transmitted using a local clock. FDDI is not synchronous but is plesiochronous.

All data is encoded prior to transmission and this uses a 4 out of 5 group code known as 4B/5B. In this scheme, every 4-bit group (16 different combinations) is mapped onto a 5-bit code (symbol). The 5-bit symbols for 4-bit data groups are chosen such that no more than two successive zeros occur. Certain 5-bit symbols that are not used for data encoding are used instead as control symbols. The 5-bit symbols are passed through a NRZI (Non Return to Zero Inverter) which produces a signal transition for a logic 1 and no change for a logic 0. The 4B/5B encoding combined with NZRI modulation guarantees that there is one signal transition at least for every three bits transmitted.

The control symbols are abbreviated to characters I, H, Q, J, K, T, R, S and L. Control symbols I, H and Q give the fibre state: I = Idle = 11111, H = Halt = 00100, Q = Quiet = 00000. Symbols J, K and T are used as frame delimiters. Symbols R and S are logical indicators ('0', '1'), and L is reserved for FDDI version II.

In an idle FDDI system, a 'token' is transmitted around the ring continuously. Each station receives the token and then re-transmits it into the ring. When a station wishes to transmit a data packet, it must 'capture' the token first. A token is captured when it has been received by a station, but not re-transmitted. Once a station holds the token it has permission to transmit data packets. After the data packets are transmitted into the FDDI ring, the station 'releases' the token by transmitting it into the ring. Further transmissions are not possible until the token is captured again.

The token format comprises a preamble, start delimiter, frame control and an end delimiter. The pre-amble (PA) contains 16 or more idle (I) symbols, which produce line changes at the maximum frequency. The start delimiter (SD) contains J and K symbols to enable the receiver to fix the correct symbol boundaries. The frame control (FC) contains 2 symbols that determine the type of information carried in the data frame and the end delimiter (ED) contains 2 T symbols.

The data frame (data packet) is shown in Figure 20.9. The header comprising PA and SD symbols, which are exactly the same as for a token. The FC symbols describe the frame type and features such as whether it is synchronous, asynchronous and the address field size. The data may contain MAC, SMT or LLC information depending on a symbol set in FC. Address information is given in the destination and source address (DA and SA), which is held in 4 or 8 symbols (as set by FC).

Figure 20.9 FDDI data frame

The data frame information field can be empty or contain an even number of symbols, up to a maximum of 9000 symbols (or 4500 bytes) including all the fields.

The data frame ends with frame check sequence (FCS) which is 8 symbols long, an end delimiter (ED), which is 1 symbol and frame status (FS), which checks the frame validity and reception. The following indicators are defined in the FS field: E = error detected, A = address recognized, and C = frame copied. Other symbols may be added possibly followed by a T symbol.

FDDI II is an extension of FDDI to support isochronous traffic. Isochronous service is required when there are strict timing constraints, such as multimedia traffic where data, digitized sound, graphics and video are integrated. Two modes of operation are supported: Basic, which is FDDI, and hybrid which is FDDI plus isochronous.

Information is carried in periodic frames called cycles and one cycle is generated every 125 microseconds. At 100 Mbps a 125 microsecond cycle can carry 12 000 bits or 3125 symbols. A cycle is divided into five preamble symbols, followed by a 24 symbol (12 byte) cycle header, 24 symbols for packet type data channel dedicated packet group (DPG) and 16 wide-band channels (WBCs), each 96 bytes wide or 3072 symbols long. In total there are 3125 symbols.

The cycle header format is illustrated in Figure 20.10. It includes 2 symbols for the start delimiter (SD), 1 symbol each for the synchronization control (C1) and sequence control (C2), 2 symbols for the cycle sequence (CS), 16 symbols (P0 to P15) containing the WBC programming template and 2 symbols for the isochronous maintenance channel (IMC). The L symbol (FDDI II only) is used to ensure the uniqueness of the cycle delimiter, SD, within the cycle packet type. Data is delimited by I and L, instead of J and K.

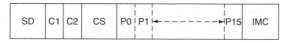

Figure 20.10   FDDI II cycle header

20.2 Wide-area networks (WANs)

If an organization has a number of separate sites distributed over a wide area, which cannot be served by a single PABX or LAN, the sites can be linked by rented or owned private lines. Provided the installations at the various sites are designed for compatible interworking, the complete system becomes a wide-area network (WAN).

Private circuits available on a rental basis from a network operator are part of the operator's bearer network, which they share with other services. Traffic sent over these rented circuits is therefore limited to the operator's standard multiplex groupings or baseband signals: depending on the way in which the rented circuit is routed, it may go through higher stages of multiplexing or even analogue-to-digital conversion and back, before emerging as a private circuit in its original form. The circuit maintains its identity as a dedicated circuit through the bearer network for the exclusive use of the renting party. The principal circuits available are:

(a) Speech-frequency channel: an analogue 300–3400 Hz channel suitable for speech or data using modems.
(b) Digital baseband channel, 64 kbit/s, data plus, if required, one or two 8 kbit/s signalling slots (increasing bit rate to 72 or 80 kbit/s). e.g. BT Kilostream.
(c) Primary digital multiplex, 2.048 Mbit/s, having 32 time slots, with slots 0 and 16 used for synchronizing and signalling, and the

remaining 30 slots for 30 digital baseband 64 kbit/s channels. e.g. BT Megastream.

The linking of LANs via rented circuits is less satisfactory, since the information circulating in a LAN has a short packet format, and differs from the baseband or multiplex signals required for the standard private circuits. A more economic solution to the linking of distant sites could be to use the packet switched public data network (PSPDN), via the X25 interface.

The standard baseband and multiplex formats and the PSPDN, are part of the international network, so the links between sites enable WANs to extend beyond national boundaries.

20.2.1 *WANs using Frame Relay*

The Frame Relay protocol operates in a similar way to X.25, but uses the ISDN as a bearer. ISDN circuits are inherently reliable and the error checking and correcting protocols in X.25 cause a high overhead in data throughput. The Frame Relay protocol only has error checking – if a frame is found to be in error it is discarded and a repeat transmission is made. Also, ISDN circuits keep user information and signalling separate; this is not compatible with X.25 but it is with Frame Relay.

The user can send between 1 and 260 octets (bytes) in each packet. There is a 2 byte address header and a 2 byte frame check sequence (FCS) at the end.

20.2.2 *Asynchronous transfer mode (ATM)*

The principle of ATM is to send data in fixed sized packets called cells. Unlike Frame Relay, each packet (cell) is exactly 53 octets (bytes) long. There are 48 bytes of user data and a 5 byte header. This payload has been designed to handle all types of traffic, such as voice, video and computer data, and as such its application is not limited to LANs. Whereas a long cell requires a lower proportion of header bytes, a short cell is less likely to cause blocking of cells from other sources. ATM is usually carried by high speed SONET or SDH networks (see Chapter 21).

20.3 Wireless connections

To allow simpler provision and movement of LAN-based equipment, wireless connections are becoming popular. Wireless LANs (IEEE

802.11), Bluetooth and the higher speed HIPERLAN are all described here. The use of infrared light, instead of radio signals, is also discussed.

Bluetooth, Wireless LAN and HIPERLAN nodes communicate through radio-based devices, which are mostly plug-in cards (ISA, PCMCIA) fitted into personal computers. The radio device has two main functions: the radio modem and the media access controller (MAC).

The radio modem modulates the data onto the radio frequency carrier and thus transmits radio signals. It also receives and demodulates transmissions from other modems. It is composed of antenna(s), amplifiers, frequency synthesizers and filters. The carrier frequency, bandwidth and transmit power are all controlled by the appropriate standard.

To overcome noise and to increase the reliability of wireless LAN systems, diversity in terms of frequency, space or time is used. Spread spectrum is a form of frequency diversity because it uses more bandwidth than necessary to avoid noisy parts of the spectrum. Retransmission and forward error correction (FEC) of the transmitted signals give temporal diversity. Spatial diversity in a radio system is achieved by using two or more antennas, to provide two paths for the radio signal.

20.3.1 Media access control (MAC)

The MAC is responsible for running the signalling protocol, which is also determined by the appropriate standard. The main characteristics of the MAC protocol are packet format (size, headers), channel access mechanisms and network management. The two channel access mechanisms used by the MAC protocol in wireless LAN systems are carrier sense multiple access/collision avoidance (CSMA/CA) and polling MAC.

CSMA/CA is the channel access mechanism used by most wireless LANs in the ISM bands. A channel access mechanism is the part of the protocol that specifies when to listen, when to transmit. The basic principles of CSMA/CA are listen before talk and contention. This is an asynchronous message passing mechanism (connectionless), delivering a best effort service, but no bandwidth and latency guarantee. Its main advantages are that it is suited for network protocols such as TCP/IP, adapts quite well to variable traffic conditions and is quite robust against interference.

CSMA/CA is derived from CSMA/CD (collision detection), which lies at the heart of Ethernet MAC: the main difference between them is the carrier sense function. On a wire, the transceiver has the ability to listen whilst transmitting and hence can detect collisions. But a wireless system cannot listen on the channel whilst transmitting, since transmit and receive frequencies are the same and because the transmit level is far higher than the receive level. Therefore the wireless MAC protocol tries to avoid, instead of detecting, collisions.

The protocol starts by listening on the channel (this is called carrier sense) and, if the channel is found to be idle, sends the first packet in the transmit queue. If it is busy, due to either another node transmission or interference, the node waits until the end of the current transmission and then starts the contention (waits a random amount of time). When its contention timer expires, if the channel is still idle, the node sends the packet. The node having chosen the shortest contention delay wins and transmits its packet. Because the contention is a random number, each node is given an equal chance to access the channel (on average).

A form of carrier sense used by some systems is request to send/clear to send (RTS/CTS). The RTS/CTS is a handshaking protocol: before sending a packet, the transmitter sends a RTS and waits for CTS from the receiver. The reception of CTS indicates that the receiver is able to receive the RTS, so the packet may then be transmitted (the channel is clear in its area). Any node within range of the receiver hears the CTS and so knows that a transmission is about to take place.

The RTS and CTS messages contain the size of the expected transmission, so any node listening will know how long the transmission will last. This is useful because the data transmission itself may not be heard. The use of RTS/CTS lowers the overhead of a collision on the medium, because RTS collisions are much shorter in time. If two nodes attempt to transmit in the same slot of the contention window, their RTSs collide and they have to try again. They lose an RTS instead of a whole packet.

Polling is a major channel access mechanism. The 802.11 standard offers a polling channel access mechanism (point co-ordination function) in addition to the CSMA/CA one. Polling is a mixture of both time division multiple access (TDMA) and CSMA/CA. TDMA is not used because although it is ideal for voice traffic (which is why it is used by GSM) it is not suitable for IP traffic that occurs in bursts. In the polling access scheme, the base station retains total control over the channel but the frame content is no longer fixed, allowing variable size packets to be sent. The base station sends a short 'poll packet' to trigger the transmission by the node. The node just waits to receive a poll packet and, upon reception, starts data transmission.

20.3.2 Error control

There is a higher error rate on the radio link than over a wire and this leads to packets being corrupted. Packet losses at the MAC layer cause problems for TCP, so most MAC protocols implement positive acknowledgements and MAC level retransmissions. Each time a node successfully receives a packet, it immediately sends back a short message (an ACK) to the transmitter. If the transmitter does not receive an ACK within a certain period after sending a packet, it will retransmit the message.

Some wireless LAN systems use fragmentation to increase the data throughput on an error-prone radio channel. Fragmentation involves breaking down big packets into small pieces before transmitting them. This adds some overhead, because packet headers are duplicated in every fragment. Each fragment is individually checked and retransmitted if necessary. In case of a corrupted packet being received, the node need only retransmit one small fragment, so it is faster.

20.3.3 Industrial scientific and medical (ISM) band

Wireless LANs and Bluetooth both use the industrial scientific and medical (ISM) frequency band at 2.4 GHz. The ITU has decreed that this band may be used for ISM purposes in all parts of the world. However, national regulators deal with specific licensing. In Europe, all equipment operating in these bands must comply with ETSI standard ETS 300 328. In the USA, any WLAN operating in this band must comply with FCC part 15. In Japan, compliance with MPT (Ministry of Post & Telecommunications) ordinance 79 is necessary. However, ISM bands are unlicensed, which means that a large number of other users may be using the same frequencies. The 2.4 GHz band also suffers from microwave oven radiation.

The ISM band regulations specify that spread spectrum techniques have to be used (either direct sequence or frequency hopping). These techniques spread the signal over a large bandwidth to reduce localized interference. The radio modem used for direct sequence is more complicated than the frequency hopping one, but the direct sequence method requires a simpler media access control (MAC) protocol. Frequency hopping is more resistant to interference, but direct sequence offers better performance when multi-path propagation is a problem. Frequency hopping is normally used.

The ISM band regulations limit the radio bandwidth to 1 MHz for frequency hopping systems. The available data rate can be increased by complex modulation schemes, allowing several data bits per symbol. This means that the receiver has to distinguish between a number of different symbols. To do this, the signal-to-noise ratio of the received signal has to be higher than if a simple two-symbol system were operating. Since the various standards limit the transmitter power level, the operating range is reduced.

20.3.4 Bluetooth

Bluetooth also operates in the 2.4 GHz license-exempt industrial, scientific and medical (ISM) band. Bluetooth uses frequency hopping and hops between 79 carriers spaced 1 MHz apart. Pseudo-random hop sequences are used so that each carrier frequency is used with equal probability. Gaussian minimum shift key (GMSK) modulation is used on these carriers. Compared to IEEE 802.11 wireless LANs, Bluetooth uses a very fast hop rate; 1600 hops per second. This means it stays on each frequency for a 625 µs time interval, which is known as a slot.

The Bluetooth protocol is a combination of circuit and packet switching. Slots can be reserved for synchronous packets and each packet is transmitted in a different hop frequency. The duration of a packet nominally covers a single slot, but can be extended to cover up to five slots. Bluetooth uses TDD (time division duplexing), which means that transmit and receive packets are carried in alternate slots. Bluetooth can support either an asynchronous data channel and up to three simultaneous synchronous voice channels, or a single channel which simultaneously supports asynchronous data and synchronous voice.

The Bluetooth packet format allows one packet to be transmitted in a slot. Each packet consists of an access code, a header and data payload. The access code is 72 bits long, the header is 54 bits long and the payload is of variable length; between 0 and 2745 bits long. Slots can be combined; a packet can be one, three, or five slots in length. Multi-slot packets are transmitted on the same frequency carrier, before the transmitter continues with the hop sequence. This reduces the transmission time lost in changing frequencies and reduces the control overhead (a five slot packet has only one access code and header, where before there were five).

Synchronous connection oriented (SCO) links support symmetrical, circuit-switched, point-to-point connections typically used for

voice. These links are defined on the channel by reserving two consecutive slots (forward and return slots) at fixed intervals. The fixed
interval size depends on the level of error correction required. Three
kinds of single-slot voice packets have been defined, each of which
carries voice data at 64 kbit/s. Voice is usually sent unprotected, since
the CVSD voice-encoding scheme is very resistant to bit errors. If the
interval is decreased, FEC rates of 1/3 or 2/3 can be selected.

Asynchronous connection-less (ACL) links support symmetrical
or asymmetrical, packet-switched, point-to-multi-point connections
typically used for burst data transmission. 1-slot, 3-slot and 5-slot data
packets are defined. Data can be sent either unprotected or protected
by a 2/3 FEC rate. The maximum data rates are obtained when an
unprotected 5-slot packet is used.

Type of packet	Symmetric (kbit/s)	Asymmetric (kbit/s)
1-slot (protected)	108.8	108.8 / 108.8
1-slot (unprotected)	172.8	172.8 / 172.8
3-slot (protected)	256.0	384.0 / 54.4
3-slot (unprotected)	384.0	576.0 / 86.4
5-slot (protected)	286.7	477.8 / 36.3
5-slot (unprotected)	432.6	721.0 (max) / 57.6

The packet definitions have been kept flexible as to whether or not
to use FEC in the payload. The packet header is always protected by a
1/3 rate FEC, this is because it contains valuable link information that
needs to survive bit errors. For data transmission, an ARQ scheme is
applied.

20.3.5 The 5 GHz band (HIPERLAN and IEEE 802.11)

HIPERLAN and satellite systems use the 5 GHz band. The band from
5.15 to 5.25 GHz (three radio channels) is available across Europe,
with 5.25 to 5.35 GHz (two extra channels) also available in some
countries, but not in the UK. These bands may be used *indoors* by both
HIPERLAN/1 and HIPERLAN/2, and transmitted power is limited to
200 mW.

HIPERLAN/2 systems use the band from 5470–5725 MHz, both
indoors and outdoors, although transmitted power is limited to 1 W.
In order to co-exist with satellite feeder links, HIPERLAN/2 systems using this band must incorporate power control and dynamic
frequency selection. HIPERLAN/1 systems do not have these facilities and therefore cannot be used here since they would risk causing
interference to satellite systems.

254

In the USA, three U-NII bands are specified. These have very liberal rules – spread spectrum is not mandated. No channels have been allocated and there are different power maximums, depending on the band being used. The low band covers 5.15 to 5.25 GHz, the mid band covers 5.25 to 5.35 GHz and the high band covers 5.725 to 5.825 GHz.

In Japan, 5.725–5.875 GHz is set aside for ISM applications, such as wireless LANs.

In the 5 GHz band, higher speeds are possible because of the availability of more bandwidth. This is typically 10 to 40 Mb/s (which in theory is also available in the 2.4 GHz band). The disadvantage with using higher frequencies is a reduced range and increased sensitivity to obstacles.

Table 20.1   PHY modes of 802.11 and HIPER-LAN/2

Modulation code	Rate	Net rate
BPSK	1/2	6 Mbps
BPSK	3/4	9 Mbps
QPSK	1/2	12 Mbps
QPSK	3/4	18 Mbps
16-QAM	3/4	36 Mbps
HIPERLAN/2 only		
64-QAM	2/3	48 Mbps
IEEE 802.11 only		
64-QAM	3/4	54 Mbps

Both the IEEE and ETSI standardization bodies have worked together in order to harmonize the physical layer for 5 GHz. The PHY layer offers the transmitting and receiving service on the wireless medium. It uses orthogonal frequency division multiplexing (OFDM) with 48 active sub-carrier plus 4 sub-carrier for pilot symbols using an FFT size of 64. The operating frequency is between 5 and 6 GHz with a bandwidth of 20 MHz per frequency channel.

OFDM does not use a single carrier nor employ frequency hopping nor use a spreading code. Instead, it simultaneously uses a large number of narrow carriers (e.g. 48) in a radio channel (20 MHz). The data is divided into several interleaved, parallel bit-streams, and each one of these bit streams modulates a separate sub-carrier. Each sub-carrier can be modulated using BPSK, QPSK, or QAM. These sub-carriers all are used for one transmission link between a mobile and an access point. One of the benefits of OFDM is the robustness against the adverse effects of multi-path propagation, common in cluttered indoor environments.

20.3.6 *Infrared*

The IEEE 802.11 provides for an infrared (IR) physical layer. Instead of a radio channel, this uses infrared light at a wavelength of 850–950 nm. The light source is an LED, which is safety rated as Class 1 (eye safe). Data modulates the LED using pulse position modulation (PPM) and achieves data transmission rates of 1 or 2 Mbps.

Infrared is intended for indoor environments, with a typical range of 10–20 m (in favourable conditions). The light is not a focussed beam, but instead is diffuse with reflections off walls and ceilings, so that line-of-sight is not required. However, a cell is limited to a single room because IR will not penetrate walls and is attenuated by glass.

21 Multiplexing

Multiplexing is the process of transmitting two or more signals over the same path without interaction. This can be achieved by separating the signals in time or frequency, or by coding the signal so that only the intended recipient can receive it. Space separation (e.g. using different pairs in a cable) is not considered to be multiplexing, since each pair provides a separate path.

Frequency division multiplexing (FDM) used to be a common technique for sharing analogue trunk circuits between 12 (or multiples of 12) separate channels. This is an analogue technique that has been superseded by the use of digital trunk circuits, where digital multiplexing is employed. FDM is still used on satellite and microwave links, although many of these are now using digital techniques. Cellular telephones use separate transmit and receive frequencies.

Time division multiplexing (TDM) is a method of interleaving digital signals from a number of channels onto one circuit. A simple example is illustrated in *Figure 21.1*, where four 600 bit/s channels are multiplexed onto one 2400 bit/s circuit. Both ends of the system must be synchronized to ensure that the data on one channel input reaches the correct channel output at the far end. In practice the circuit's data rate would have to be higher than 2400 bit/s to allow for synchronizing data to be sent.

This chapter will concentrate on TDM systems that allocate a fixed time slot for each channel, and will cover Plesiosynchronous Digital Hierarchy (PDH), Synchronous Digital Hierarchy (SDH) and Synchronous Optical NETwork (SONET). Statistical multiplexing can be provided by packet transmission, such as Frame Relay and ATM, and these methods will be discussed in Chapter 22.

Optical Wavelength Division Multiplexing (WDM) is the same principle as FDM. However, the optical techniques differ in many respects and are still evolving. In electrical FDM only a single line driver (power amplifier) is required because signals on the different channels have been modulated onto separate carriers. In optical WDM each channel requires a separate source (laser) that produces a single wavelength. Experimental systems that combine 100 wavelengths have been built, but commercial equipment is already available for combining 16 wavelengths.

Digital signals can be multiplexed by coding each source individually. Instead of transmitting logical 1s and 0s, a data pattern is

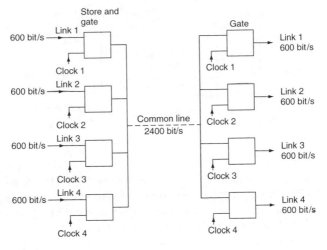

Figure 21.1

assigned to each bit. The pattern for a logical 1 could be a ten bit word (e.g. 1 011 010 011), a separate pattern would represent a logical 0. Correlation (i.e. pattern matching) between a known signal and a signal received with a certain code gives an indication of whether the bit received is a 1 or a 0. This technique allows the occasional error in the data path without affecting the received bit. More often, it allows several users to transmit at the same frequency. Each user is allocated a different data pattern so that only the correct person will be able to decode the signal; all other signals will be ignored.

21.1 Frequency division multiplex – basic group

A modulated carrier signal was used for many years as the transmission medium along copper wires between telephone exchanges. Amplitude modulation by the voice frequency has been internationally adopted, and to conserve bandwidth the carrier is suppressed and only one sideband transmitted per voice channel. This is usually the lower sideband, at least in the UK.

At the receiving station, a locally generated carrier of the same frequency as the suppressed carrier is used to recover the original voice signal. Clearly any frequency difference between transmitter and receiver carrier frequencies will cause a frequency shift of the analogue voice waveform; in practice a 1 Hz shift is the typical allowed error. *Figure 21.2* illustrates the process.

258

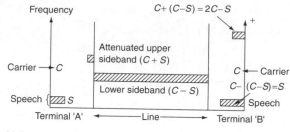

Figure 21.2

The ITU-T defined basic group consists of 12, 4 kHz channels, each one of which will be modulated as described on a quartz crystal oscillator. The 12 crystal frequencies are spaced 4 kHz apart and occupy the band 60–108 kHz. The reason for this range of frequencies is that cable attenuation versus frequency is almost flat.

The ITU-T (formerly CCITT) *Yellow Book,* Vol. 3, Rec. G232, fully describes the tolerance limits of attenuation versus frequency for each band, the interchannel cross-talk allowed and the permitted level of residual carrier frequency. These requirements are met by stringent design of post-modulator channel filters, which normally employ quartz crystal elements to achieve the desired attenuation slope. The audio signal is also normally filtered before modulation to restrict its bandwidth to 0.3–3.4 kHz; either traditional *LC* or active filters may be used. Some systems may use out-band signalling, for example, at 3825 Hz, but this clearly demands an even greater attenuation slope in the channel separation filters.

Figure 21.3 shows a typical mask for attenuation versus frequency for the voice channel of 12 channel equipment, and *Figure 21.4* shows the basic group.

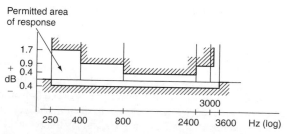

Figure 21.3

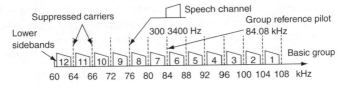

Figure 21.4

21.1.1 FDM supergroup

Just as the 12 voice channels are used to modulate carriers in a basic group, so are five basic groups, each assembled from different voice channels, used to amplitude-modulate higher-frequency carriers at 420 kHz, 516 kHz, 564 kHz and 612 kHz. As before, the lower sidebands are filtered out and combined to occupy a band 312–552 kHz. This supergroup then modulates a carrier at 564 kHz and the lower sideband, extending from 12 kHz to 252 kHz, is transmitted to line, containing 60 voice channels in its 240 kHz bandwidth. *Figure 21.5* shows the assembly containing 60 voice channels. The pilot is at 411.92 kHz.

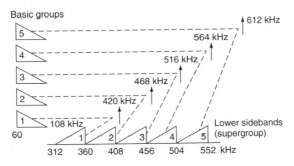

Figure 21.5

21.1.2 FDM hypergroups (or master groups)

In some countries a master or hypergroup consists of five 60 channel supergroups modulated up to 812–2204 kHz, and three such groups are then assembled to form a 900 channel supermaster group from 8516 to 12 388 kHz.

Although it is technically feasible to extend multiplexing to even higher frequencies, for example a 10 800 channel 60 MHz system is in use in some European countries, the introduction of digital transmission techniques has increasingly eliminated any commercial reason for FDM development. The 12 MHz channel has now been improved to carry 3600 channels by extension to 18 MHz and this is likely to be the last advance in the FDM system. A 60 MHz FDM network consists of 12×900 circuit hypergroups contained within a line spectrum of 4–60 MHz.

21.2 Time division multiplex 64 kbit/s baseband signal

The digital baseband signal adopted in Europe and North America is at a 64 kbit/s rate. A 4 kHz speech channel is converted into digitized samples; each sample is allocated an 8 bit code corresponding to one of 256 (2^8) quantizing levels, and the sampling rate is 8 kHz. The same 8 bit word or octet may also be obtained from the IA5 alphabet (7 bits) plus a parity bit. A signalling bit is sometimes added to each octet (increasing the rate to 72 kbit/s), but this is normally removed at the multiplexing stage to restore the bit rate to 64 kbit/s, the signalling information being transferred to the signalling channel in the multiplex stream.

The bit duration in the 30 channel PCM system is 488 ns, so an octet occupies $8 \times 0.488 = 3.9$ μs. The interval between successive samples of a voice channel is 125 μs (i.e. 1/8000), so each sample occupies only a small fraction of the sampling time, which in this context is called a frame. The remaining time within the 125 μs frame is allocated to other voice channels as described in the next section.

21.2.1 European Level 1 (primary) multiplex 2048 kbit/s (E1)

The previous section describes how a 125 μs frame contains an 8 bit amplitude sample, lasting 3.9 μs, from a voice channel. In time division multiplexing the remainder of the frame contains digitized samples from other voice channels, and since $125 \div 3.9 = 32$, that is the total number of channels that can be contained in one frame. Each frame is therefore divided into 32 time slots (TS) 30 of which are used for voice channels and 2 for control purposes. The time slots are designated TSO–TS31; TSO contains a frame-alignment signal, TS16 is reserved for interexchange signalling.

Each TS contains 8 binary digits, and so the gross digit rate is $8000 \times 8 \times 32 = 2048$ kbit/s.

The 30 speech channels are interleaved sequentially, sample by sample in each 125 μs frame, but the TSO frame-alignment word alternates between two values, and the TS16 slot, which has to carry signalling information for all 30 channels, is further submultiplexed over a period of 16 frames (i.e. 2 ms), this group of 16 frames being called a multiframe.

The frame-alignment signal in TSO is 0 011 011 transmitted in binary digit spaces 2–8 in alternate frames. Digit space 1 of TSO is reserved for international use, but is normally set to a '1'. In the intermediary frames only binary digit space 2 is fixed as a '1'; the other spaces are not allocated except that binary space 3 indicates the state of the multiplex alarms. Likewise, binary space 6 of TS16 in frame 0 of a multiframe indicates the state of alarms associated with channel signalling.

Figure 21.6 shows the 30 channel frame format.

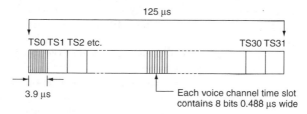

Figure 21.6 TSO, TS16 are used for signalling; TS1–TS15 and TS17–TS31 are used for voice channels

The actual signals sent to line are not unipolar since such a pulse train would contain a large DC component (e.g. a succession of all '1's) together with high amplitude low-frequency components which are undesirable. The 2048 kbit/s binary digit stream is instead encoded to HDB3 (see Section 21.2.3).

The 30 channel 2048 kbit/s stream is the basic building module of the digital network in Europe. It is the bit rate at which switching takes place in stored programme control (SPC) exchanges and is the format of the BT Megastream data links. The multiplex is widely adopted as the access link from digital PABXs and LANs to SPC exchanges, and to interconnect PABXs or LANs on separated sites.

21.2.2 North American Level 1 (primary) multiplex 1544 kbit/s (T1)

The 24 channel PCM multiplex used in North America is similar except that the line bit rate is $24 \times 8 \times 8000 + 8000 = 1544\text{kbit/s}$ and the transmitted bit stream uses AMI instead of HDB3 encoding (see Section 21.2.3).

Time slots (TS), or channels, in the T1 system are numbered 1 to 24; there are 23 available to carry traffic, TS24 is used for synchronization. Each time slot carries 8 bits, sampled at 8 kHz. An additional bit is sent at the end of each frame (the frame or 'F' bit) so there are 24 8-bit time slots, plus 1 bit = 193 bits per frame.

The synchronization word is coded as 1 0 1 1 1 Y R 0. The Y and R are dependent upon the data multiplexer and are for the manufacturer's use.

The 'F' bit is used to identify the frame. A six bit data pattern of 1 0 0 0 1 1 is transmitted, followed by the inverse: 0 1 1 1 0 0. The complete 12 frame sequence is known as a superframe. The use of an 'F' bit is necessary because signalling information is transmitted using 'stolen' bits from each channel, rather than using a separate time slot as in the E1 system. Bits are not stolen from time slots in every frame, because this would degrade the quality of a voice channel. Instead, the least significant bit (the eighth bit) of every sixth frame is used for signalling.

There are two virtual signalling channels (A and B) created by the use of stolen bits. Signalling Channel A is derived from bits stolen from time slots during the sixth frame. Signalling Channel B is derived from bits stolen from time slots during the twelfth frame. This is repeated for every superframe. Each channel has a data rate of 24 bits per 12 frames (1.5 ms), or 16 kbit/s.

21.2.3 Alternate mark inversion and high-density bipolar

To prevent a net d.c. content, the digital signals sent to line in TDM systems are not unipolar; a two-fold process of encoding is used to produce a bipolar signal which cannot contain a long string of successive '0's that might cause loss of synchronization.

Figure 21.7 shows the process of alternate mark inversion (AMI) in which alternate bits '0's and '1's are first inverted in sense, that is a '1' becomes a '0'; and vice versa. Then alternate '1's are changed in level polarity to produce a bipolar waveform.

In order to avoid any problem caused by a long succession of '0's a '1' bit is deliberately introduced after a given number of '0's. In the 30 channel PCM structure this '1' is introduced after three successive '0's and the code is then referred to as high-density bipolar-3 (HDB3) (see ITU-T Rec. G703).

The result of using HDB3 is that in any five-digit time slot a minimum of two '1's will occur thus ensuring adequate clock rate recovery in a digital regenerator. The '1' deliberately introduced after three successive '0's is allocated the same polarity as the previous

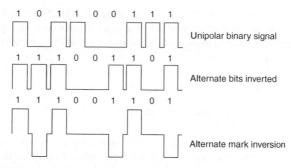

Figure 21.7

mark, but successive violations like this are transmitted in opposite polarity. Even so an even number of '1's between violations would result in successive violations being of the same polarity. To obviate this the first '0' of a sequence is changed to a '1' and called the parity bit, while the fourth '0' is changed to a '1' which violates the parity bit. This is illustrated in *Figure 21.8*.

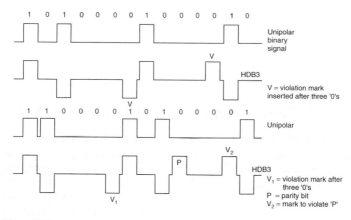

Figure 21.8

The digital signal at the coaxial output port to the line system is, in accordance with ITU-T Recommendation G703, at a peak voltage of 2.37 volts with 50% duty cycle and nominally rectangular pulse shape. A zero is sent as zero voltage. The power density distribution for an HDB3 encoded random binary signal peaks over the range 0.8 to 1.2 MHz in the 2048 kbit/s system.

Modern large-scale integration technology in CMOS has produced single-chip circuitry capable of simultaneous synchronous coding and decoding of HDB3 signals and, furthermore, provides error monitoring to detect code violations.

21.2.4 Coded mark inversion (CMI)

The highest rate 68 Mbit/s and 140 Mbit/s levels of the TDM hierarchy may use an interface code called coded mark inversion (CMI). In this two-level, non-return to zero (NRZ), code binary '1's are allocated alternate positive and negative levels for the whole digit time slot, but binary '0' is represented by a negative level followed by a positive level, each lasting for one-half the duration of a digit time slot.

Figure 21.9 illustrates the principle. CMI contains a high level of signal at digital clock rate, which assists system synchronization and is also relatively easy to implement in hardware.

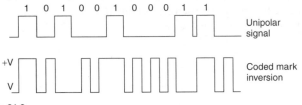

Figure 21.9

21.3 Plesiosynchronous digital hierarchy (PDH)

The 2 Mbit/s 30 channel PCM is used as the basic building module of a digital transmission system currently operating in Europe at 2, 8, 34, 120 and 140 Mbit/s. It is now possible to combine four 140 Mbit/s tributaries into a 565 Mbit/s highway over monomode optical fibre. In the second stage of this hierarchy the 8 Mbit/s signal is obtained by interleaving on a bit-by-bit basis the outputs of four separate 2 Mbit/s channels. The equipment which does this is called a digital muldex – muldex being a contraction of multiplex-demultiplex, and thus is descriptive of both directions of working. The inputs/outputs on the low-frequency side of a muldex are generally called the tributaries, the term muldex or MUX inputs/outputs being reserved for the high-frequency side.

Figure 21.10 (a) shows the interconnections of a 30 channel PCM to a 2–8 Mbit/s muldex and also how the four tributary data streams

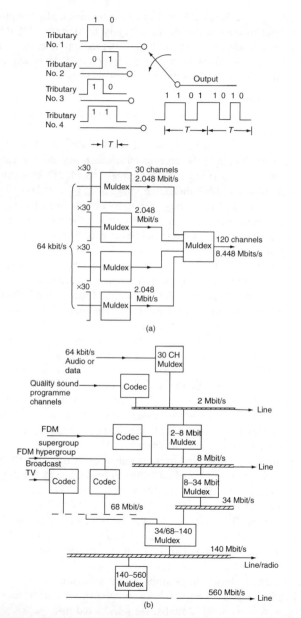

Figure 21.10

combine on a timeshare basis as they are each sequentially sampled for a time equal to one-quarter of the tributary bit duration. HDB3 line signals are decoded to unipolar binary before multiplexing. *Figure 21.10(b)* illustrates the assembly of the complete PCM structure.

The main functions of a muldex are as follows:

(a) *Transmitting*:

1 Combine bits from the four tributaries, after decoding HDB3 to binary.

2 Insert a 10 bit frame-alignment word (FAW) to allow accurate demultiplexing at the receive terminal. In the present case the frame is subdivided into four equal sets of 212 bits. At a bit rate of 8448 kbit/s the frame duration is $(4 \times 212) \div (8448 \times 1000) = 100.4$ µs, corresponding to a frame rate of 9.962 kHz. The frame starts with the FAW, and at the receiving end the function of all other digits can be checked by their displacement from the FAW.

3 Perform JUSTIFICATION – the process by which four tributaries running at marginally different rates are interleaved. The permitted tolerance on the frequency of the 2 Mbit/s channel is 50 parts per million or ±102 bit/s. The four tributaries are said to be PLESIOCHRONOUS and since the 2–8 Mbit/s muldex cannot control their rates it has effectively to synchronize to all four inputs. This it does by using a data store following the HDB3 decoder in which data is entered at the line bit rate (nominally 2048 kbit/s) and read out at a nominal 2112 kHz generated by a master clock in the muldex, but this 2112 kHz signal has bits removed at defined points to make its average rate equal that of the write clock (i.e. recovered 2048 kbit/s) over a period of several frames.

The process is controlled by the content of justification time slots (JDT) occupying slots 5–8 of set 4 in each frame, and by a justification control word (JCW). This is a 3 bit word (000 or 111) occupying bits 1, 2, 3 and 4 of sets 2, 3, and 4. When the tributary bit rate is exactly 2048 kbit/s the JDT carries a '1' for every other frame, on average (i.e. 50%); if the input rate is high it is used more often, and if low, less often. The JCW is 111 if the frame is 'justified' that is does not carry a valid tributary bit at JDT.

Figure 21.11 shows the 8 Mbit/s frame structure.

When justification is successfully completed, the four tributary signals, now running at 2112 kbit/s, are interleaved into one 8448 kbit/s stream.

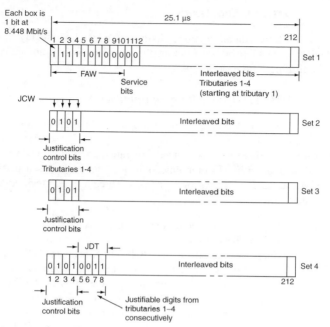

Figure 21.11 Note: the JCWs shown are 000, 111, 000 and 111 for tributaries 1–4 respectively

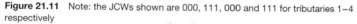

    4 Transmit an alarm signal (bits 11 and 12) should system failure occur.
    5 Recode the interleaved signal, FAW, etc. into HDB3 before transmission.
(b) *Receiving*:
    1 Demultiplex the input high rate signal into four tributaries at 2048 kbit/s.
    2 Extract and monitor the FAW.
    3 Interpret JCW and JDT for synchronization.
    4 Display alarms.

21.3.1 TDM 8–34, 34–140 and 8–120 Mbit/s

A similar process of interleaving frame alignment and justification to that used in the 2–8 Mbit/s muldex produces these higher-rate systems. However, HDB3 is not used at the higher rates where, instead, coded mark inversion (CMI) is easier to implement in hardware (see Section 21.2.3)

268

21.3.2 TDM 140–560 Mbit/s

Four 140 Mbit/s tributaries are digitally multiplexed as already described and the high bit rate output then used to switch an optical
source feeding a fibre-optic link.

21.3.3 FDM–TDM transition equipment

During the period of change from analogue FDM transmission systems to digital TDM systems, which is likely to take several years
internationally, there is a clear need for transition equipment, which
will allow uninterrupted communication over links composed partly
of FDM and partly of TDM transmissions. Three different techniques
have been proposed:

1 *Transmultiplexers*: This equipment transforms the 60 channels
  assembled on to a basic supergroup to provide direct interconnection
  to 60 channels assembled on two 2048 kbit/s primary TDM data
  streams, so as, for example, to provide two-way communication
  between analogue and digital switchers (exchanges) in the PSTN
  or in a private network. *Figure 21.12* shows the block diagram of
  a transmultiplexer which uses digital signal processing to filter and
  multiplex the TDM signals before final conversion to an FDM signal. The signals on the digital side are sampled at $8N$ kHz, where
  N is an integer greater than the number of channels in the FDM
  multiplex.

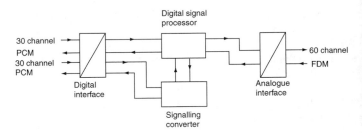

Figure 21.12

2 *FDM codecs*: It is proposed that, just as a voice channel can be digitized for transmission as PCM, so could a basic supergroup or even a
  15 supergroup assembly be transformed via a suitable codec (code-
  decode circuit), based on digital/analogue and analogue/digital convertors. As with voice-channel codecs, consideration of the amount

of quantizing noise and timing jitter introduced by the processes is important. Compared with transmultiplexers, FDM codecs use simpler circuits which may be implemented with commercial integrated circuits and thus will almost certainly be cheaper. A codec cannot carry as many voice channels as an all-digital muldex (see Section 19.6) operating at the same bit rate, but, since it encodes the analogue multichannel in its entirety, a codec can deal equally with voice, programme and data channels whereas a transmultiplexer is restricted to 4 kHz voice channels. A simplified block diagram of a codec is shown in *Figure 21.13*.

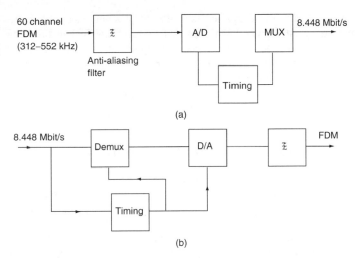

Figure 21.13

3 *Data-in-voice and data-over-voice modems*: This equipment enables a digital data stream to be carried over an analogue transmission system. For example, a 2048 kbit/s signal could be transmitted over the bandwidth normally allocated to two adjacent supergroups within a hypergroup, that is 1804–2292 kHz, and this is described as a DIV (Data in Voice).

In DIV modems, the digital stream is carried by that normally unused part of an FDM assembly above the highest frequency arising from FDM modulation, and above any system pilot frequency. In some systems this may not be possible because of poor noise performance (even for digital signals) or existing use by supervisory signals, and the technique has found little practical implementation.

Data-over-voice (DOV) modems transmit data above the voice band. This is an FDM technique since data and voice use separate frequency bands. Fujitsu have produced a modem that takes data on an E1 link and modulates a carrier in the 1804 to 2292 kHz band. This band is above the basic FDM supergroup frequency range and can thus be carried over the same circuit. DOV can also be applied to local access networks by using a modem with a much higher than usual carrier frequency.

21.3.4 Multiplexing time delay

Section 21.2.1 has explained how the 30 channel system is multiplexed by interleaving 8 bit words from each channel in turn into sequential time slots. Data on each tributary (data of course includes quantized speech) is stored in a buffer register and then transmitted in serial form to the high-speed line during the time slot period allocated to that tributary. There is thus inevitably a delay period before a complete character is transmitted.

21.4 Synchronous digital hierarchy

One of the disadvantages of the existing plesiochronous digital multiplex system is that a 140 Mbit/s stream, for example, has to be demultiplexed all the way down the chain in order to identify any of the 64×2 Mbit/s tributaries.

ITU-T Recommendations G707–G709 describe a synchronous digital hierarchy (SDH) with higher-order bit rates of 155.52 Mbit/s and 622.08 Mbit/s and which allow 'jump multiplexing' from 2 Mbit/s to 155 Mbit/s; 'drop and insert' whereby any 2 Mbit/s stream can be added to or subtracted from a 155 Mbit muldex; and the facility for direct interconnection of any 2 Mbit tributary between transmission systems without demultiplexing.

The advantage of SDH over PDH is that both the cost and size of the multiplexing equipment is reduced. In an SDH system a single shelf in a rack can be used to convert directly from 155 to 2 Mbit/s. A PDH multiplexing system requires one or more racks of equipment. A 140 Mbit/s stream is first converted to 34 Mbit/s, then to 8 Mbit/s and finally to 2 Mbit/s before channels can be extracted or inserted. The reverse process is then applied to restore the data rate to 140 Mbit/s for subsequent re-transmission. Since far less equipment (albeit more sophisticated) is required, the cost is reduced.

The SDH data rates are compatible with the USA's Synchronous Optical NETwork (SONET). The ITU-T has defined the basic SDH data rate of 155.52 Mbit/s and this is known as Synchronous Transport Module number 1 (STM-1). Multiples of this rate are numbered STM-n, where n is the multiplier. Thus STM-4 has a data rate of 622.08 Mbit/s.

The base rate of 155.52 Mbit/s makes SDH suitable for carrying data from multiples of DS0 (64 kbit/s), E1, E2, E3 and E4 systems. A dual optical fibre ring connects optical multiplexors together. One fibre carries the working copy of the data and one carries a protected copy. Should the quality of the working copy be below standard, the protected copy will be used and an alarm will warn of an impending optical system failure. Output from the multiplexor is at STM-n data rates and requires a terminal adaptor to extract E1, E2, E3 or E4 data streams.

21.5 Synchronous optical network (SONET)

The SONET data rates have been defined by ANSI, the basic rate being 51.84 Mbit/s and known as Synchronous Transport Signal number 1 (STS-1). Multiples of this base rate are numbered STS-n, where n is the multiplier. The SDH base rate of 155.52 Mbit/s (STM-1) is equal to the SONET data rate of STS-3.

The 51.84 Mbit/s base rate makes SONET suitable for carrying data from multiples of DS0 (64 kbit/s), DS1 (T1), DS2 (T2) and DS3 (T3) systems. Like the SDH network, SONET uses a dual optical fibre ring to connect optical multiplexors together, with one fibre used for protection against failure of the other. Output from the multiplexor is at STS-n data rates and requires a terminal adaptor to extract T1, T2 or T3 data streams. A digital cross connect switch is used to link to other SONET networks, rather like a bridge linking two Ethernet LANs.

The optical network is defined in terms of optical carrier rate (OC-n). The OC rate is aligned with the STS-n rates, thus OC-1 has a data rate of 51.84 Mbit/s. The relationship between the OC, SONET and SDH rates are given in *Table 21.1*.

Apart from the difference between SONET and SDH base rates, all rates are equal which allows STM-n and STS-n data streams to be multiplexed onto the same fibre network. Monomode fibre is normally used for telecommunications applications, with lasers operating at either 1300 or 1550 nm wavelength. Multimode fibre can be used

Table 21.1 OC/SONET/SDH transmission rates

Optical system	SONET	SDH	Rate (Mbit/s)
OC-1	STS-1		51.84
OC-3	STS-3	STM-1	155.52
OC-9	STS-9	STM-3	466.56
OC-12	STS-12	STM-4	622.08
OC-18	STS-18	STM-6	933.12
OC-24	STS-24	STM-8	1244.16
OC-36	STS-36	STM-12	1866.23
OC-48	STS-48	STM-16	2488.32
OC-96	STS-96	STM-32	4876.64
OC-192	STS-192	STM-64	9953.28

at the lower bit rates (up to OC-12, or 622.08 Mbit/s) over short distances such as on a university campus.

21.6 Wavelength division multiplexing (WDM)

Wavelength division multiplexing allows two or more optical systems to operate over the same fibre. This may be required to increase the capacity of an optical fibre route, or could be used to allow bi-directional transmission over a single fibre. A 16-channel system used in conjunction with SONET systems has produced an enormous data carrying capacity. A 4-channel system combined with STS-48 SONET systems has the same capacity as an STS-192 system (about 10 Gbit/s), but is cost competitive and may be preferred in some applications.

Separating optical wavelengths is similar to separating the frequencies of electrical signals. A filter is needed which passes the desired wavelength but blocks others. Optical filters can be made from defraction gratings, which behave like prisms; these have a series a parallel ridges in the optical medium (plastic or glass) which defract light at an angle which is wavelength dependent. Multiple refractions may be necessary to provide the required separation, rather like multiple stages in an electrical filter.

If the two wavelengths are in separate 'windows' (i.e. 1300 and 1550 nm) the requirements on the filter are not too demanding. The problem is far greater when both wavelengths operate in the same window. This is because the closer the wavelengths are to each other, the harder they are to separate. Precision mechanical and optical engineering has produced a commercial 16-channel system with all channels operating in the 1550 nm window. Research into obtaining more channels has so far produced a 100-channel system.

22 Packet switched data

This chapter deals with the format of packet data and describes how packets are switched through the network. The application is primarily wide area networks, although not necessarily since some packet orientated services can carry voice and video traffic as well as data.

Packet data is often transmitted in bursts of a pre-defined length. To transmit a file, the data is broken down into segments. The data segments are then made into packets by adding a head and (possibly) a tail. Start and address data is added at the beginning of the segment. Error checking data and a stop byte may be added at the end. Each packet is then a short self-contained burst of data. A Frame Relay packet or 'frame' is illustrated in *Figure 22.1*.

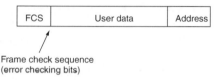

Frame check sequence
(error checking bits)

Figure 22.1 A Frame Relay 'frame'

Packet data is transmitted over networks to packet switches. A packet switch looks at the address header and re-transmits the entire packet to that address. If the most direct route is busy, a secondary route will be used. Using multiple routes can result in packets arriving out of order, and some protocols allow for this by storing and re-ordering the packets as they arrive.

Some protocols allow for receiving stations to operate at a slower speed to the transmitting station. This is achieved by numbering the packets and waiting for acknowledgements; a 'sliding window' is used to limit the number of packets sent but not acknowledged. The sliding window contains a number that is the difference between the number of packets sent and the number of packets acknowledged. As the acknowledged packet number rises, the window slides to allow a higher transmit packet number. This concept is illustrated in *Figure 22.2*.

22.1 Packet switching

Before describing the data structure, it is necessary to describe how the packets are switched through the network. An address header

274

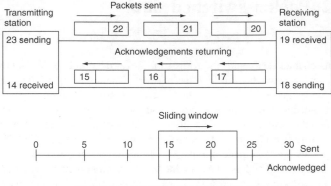

24 cannot be sent until acknowledgement of 15 is received

Figure 22.2 Sliding window concept

precedes the data, and this is used to find the addressee (rather like how a telephone number is used to route a call). Nodes along the network examine this header to determine where the packet should be routed.

The address may not be exactly like a telephone number. Protocols such as X.25, Frame Relay and asynchronous transfer mode (ATM) only have a short address field. For these protocols it is necessary to provide a temporary address. The route is determined at set-up time and all packets containing the temporary address are routed through to the same end point. The actual route may depend upon network availability; if a route is heavily loaded the switch will select a different, less congested route. When the link is no longer required a message is sent to the network to free the temporary address for other users. In the switched multi-megabit data service (SMDS) the address field is long enough (60 bits) for end-to-end routing.

Packet switched networks using temporary addresses are said to be virtual circuits. A virtual circuit is one where the circuit appears to be in place, since data entering one port exits from another port. It is virtual because, when the network removes the temporary address, the circuit no longer exists.

22.1.1 Permanent virtual circuits

A permanent virtual circuit (PVC) is like a 'private wire' used in telephony. Channels are purchased to give an end-to-end data path between fixed points. Sender and receiver addresses are assigned for the two ends and no other addresses are accepted by the network.

Other users are barred from using these addresses and so the channels are secure. They are 'virtually permanent' since the route taken by cells is determined by availability. If a link fails or is busy cells will be routed via another path.

22.1.2 Switched virtual circuits

A switched virtual circuit (SVC) is like the public switched telephone network since paths can be set up between any users.

1 A 'connect request' packet is sent to the terminal to which data needs to be sent (this is equivalent to ringing of a telephone bell).
2 Acceptance of the connection requires a 'connect accepted' message to be sent back to the caller (equivalent to picking up the telephone handset). If the connection is not accepted a 'connect refused' message is sent to the caller.
3 A successful connection results in a temporary channel number being assigned by the network; this is the number that is used in the address field of the packet or cell header.

Public X.25 networks commonly use SVCs, but many private organizations prefer to use the more secure PVCs.

22.2 High-level data link control (HDLC) frame structure

The format used for X.25 is the HDLC frame. Its structure is similar to that used by Frame Relay and is illustrated in *Figure 22.3*.

The start and end flags have the data byte 0 1 1 1 1 1 1 0. The start flag is followed by an address that can be 0, 8 or 16 bits long, depending upon the frame type. The address could be that of the sender or the intended recipient. Referring to *Figure 22.3*, the data link connection identifier (DCLI) is the 'virtual address' allocated by the network. This may be only 6 bits long, so the first 6 bits are used for this. Bit 7 is the command/response (C/R) bit and bit 8 is the extended address (EA) bit.

The EA bit determines whether there is a second address byte; if it is set to '1', data follows. If the EA bit is set to '0' a second address byte follows, beginning with a further 4 bits of the virtual address. The forward error correction number (FECN), backward error correction number (BECN), the DE bit and a second EA bit complete the byte. The second EA bit is set to '1' by default, but there is an option in the standard to extend the address by a further 8 or 16 bits if required.

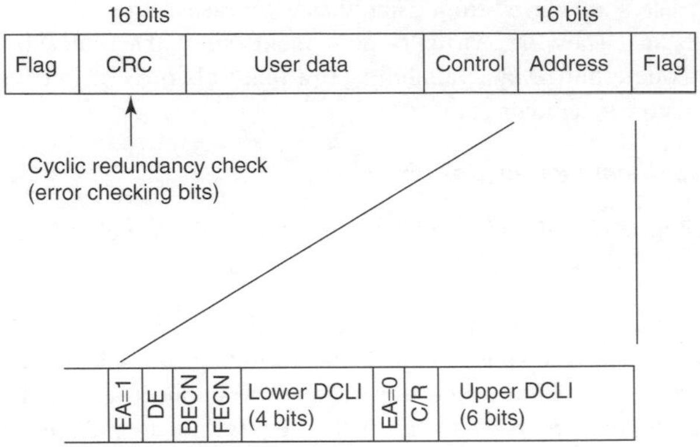

Figure 22.3 HDLC frame for X.25

Following the address is a control byte. This is used to specify whether the frame is unnumbered, supervisory, or information type. These are detailed here:

1 An unnumbered frame is used for link management. This is used to set up the relationship (modes) between terminals on the network. The normal response mode is where there is one primary and several secondary terminals. The primary terminal has to give permission for the secondary terminals to communicate. The asynchronous response mode is similar, but allows secondary terminals to communicate without permission. The asynchronous balanced mode is where two primary terminals are connected, and have equal status.

2 A supervisory frame is used for error and flow control. Each frame is numbered incrementally, so that the receiving station knows the order in which the data should be placed when removed from the packets. The reply packet contains the frame number, so the sending station knows if any packets have to be re-sent.

3 An information frame contains the data for transmission. Data acknowledgements of previous transmissions are sent with the return data on full duplex links, to save sending a separate supervisory frame.

The data transmitted does not have to be a discrete number of bytes. Any number of bits (i.e. not necessarily multiples of 8) can be transmitted.

Two bytes are used for cyclic redundancy check after the data. Error checking procedures can determine if any of the data was sent in error. Any errored packets are discarded and, when the transmitting station does not receive acknowledgement of receipt, a replacement packet is sent.

A second flag byte of 0 1 1 1 1 1 1 0 is sent to indicate the end of frame.

22.2.1 Link access procedures (LAP)

The manner in which HDLC used depends upon the transmission path. The ITU-T has developed LAP standard protocols:

- LAPB is used on the ISDN B channel (64 kbit/s). X.25 networks are based on LAPB, which ensures that data arrives correctly and in order.
- LAPD is used on the ISDN D channel (16 kbit/s) and is used for signalling.
- LAPM is used on modem circuits. It provides asynchronous to synchronous conversion, flow control and error detection.
- LAPF is used for Frame Relay.

22.2.2 X.25 networks

Networks for X.25 can be permanent or dial-up and are used for text file transfer. Permanent circuits may use a digital leased line (e.g. 64 kbit/s), or an analogue line with a modem and a packet assembler/disassembler (PAD) fitted. The PAD provides serial to parallel conversion. Dial-up circuits using a modem and a PAD can be used or, if available, the B channel of an ISDN circuit.

The HDLC frame used with the LAPB protocol provides an error free link by using the cyclic redundancy check (CRC) error checking bits. Flow control is used so that buffers in a slow receiving device do not overflow.

A store and forward protocol is used. In an X.25 path there will be several nodes where data is received and re-transmitted as it is routed through the network. If a packet is received with an error at any node, the receiving station sends a message back to the previous node, asking for re-transmission of the packet. The data is only used when it is received error free, and this makes the end-to-end transmission process quite slow.

22.2.3 Frame Relay

The error checking and flow control required on analogue telecommunication networks is less important on digital networks. The digital networks provide an almost error free transmission system. The error checking protocols in X.25 which slowed the data rate considerably are no longer required. As a result, Frame Relay provides a much faster transmission protocol.

Unlike X.25 there is no store and forward process. Error checking is only performed by the end user. All data passing through nodes in the Frame Relay path is re-transmitted without any error checking or storage. Data is transmitted to the next node whilst still reading the packet in from the previous node.

If any errors are made during transmission, the data packet has to be re-transmitted across the whole network route. However, since errors are rare, re-transmission takes place infrequently and the overall data rate is far higher than in an X.25 network.

Non-Frame Relay terminals can access the Frame Relay network using a Frame Relay Assembler Disassembler (FRAD). This accepts X.25, SNA/BSC or asynchronous data and converts into a suitable format. A Frame Relay gateway which provides similar functionality to the FRAD is often provided at a switching node, to link X.25 and Frame Relay networks.

22.3 Switched multi-megabit data service (SMDS)

The switched multi-megabit data service (SMDS) provides a connectionless packet switched service. Each packet has a 40 byte header and up to 9188 bytes (octets) of user data. The SMDS may actually use asynchronous transfer mode (ATM) cells for transmission by breaking the packet down into a 5 byte header and a 48 byte payload. The payload only contains 44 bytes of data, since 2 bytes are used as a header and 2 bytes are used as a trailer.

The important word in the title is 'service'. The transport mechanism could be ATM, as mentioned, or Frame Relay. It is connectionless because it is the underlying transport technology that provides the connection – and this could be using packets which travel through various paths.

22.4 Asynchronous transfer mode (ATM)

Any large organization may have a communications infrastructure with a number of independent networks handling voice, image and data traffic. Keeping control of all these separate networks and ensuring that efficient use is made of them is difficult. ATM provides a method to integrate these discrete networks whilst maintaining the efficiency provided by the separate networks dedicated to one type of traffic.

The advantage of a single integrated network is that information can be shared over a wide area network (WAN). Also, there are applications for more bandwidth and for identical functionality from all locations in the network. Multimedia applications, that integrate two or more media (data, voice, image or video) are becoming more common; these include videoconferencing and computer telephony. Frame Relay and SMDS cannot handle time-sensitive traffic such as real-time video and voice. The Frame Relay and SMDS protocols were developed specifically for WAN use. The advantage of ATM is that it can handle multi-media traffic and can carry information from the workstation right through the WAN.

ATM uses fixed-length cells of 53 bytes: 48 bytes of user data (payload) and 5 address bytes in a header. Before transmission, the user data (e.g. a file) is broken down into multiples of 48 bytes to form the separate payloads. A header is attached before the cell is transmitted on the ATM network. At the destination, the user data is re-assembled by concatenating (linking together in series) the cell payloads.

To reduce processing overheads (which cause delay), a 'virtual address' is used. This is short and allows the header to be a small part of the cell. The full address is used to set up a virtual path initially, but this only appears in the first cell of the packet transmission sequence. Once the ATM network has assigned a virtual address, the subsequent packets only have to contain this abbreviated destination. The use of a virtual address makes ATM 'connection-oriented', unlike SMDS. An ATM cell is illustrated in *Figure 22.4*.

A 'virtual path' is a semi-permanent connection between the end nodes – for example, from a company's head office to an ATM switch. A second virtual path may be between the ATM switch and the company's factory. A virtual path can be likened to a cable containing many circuits (in reality it is likely to be a number of time slots in a multiplexed system).

A 'virtual circuit' or 'virtual channel' is part of a virtual path. It can be likened to a pair of wires in a cable. Thus virtual circuit 1 on path 1 is different from virtual circuit 1 on path 2, in the same way that pair 1 in one cable is different from pair 1 in a second cable. The virtual circuit is provided on a temporary basis, and this can be likened to a telephone trunk circuit between central office switches: the trunk is in use while the call is in progress, but is free for others to

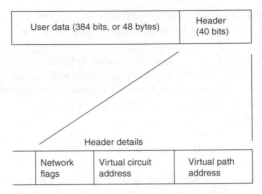

	User data (384 bits, or 48 bytes)		Header (40 bits)

Header details

Network flags	Virtual circuit address	Virtual path address

Figure 22.4 ATM cell

use when the call clears. If the connection between the two telephone users is remade after a break, a different trunk circuit is likely to be provided. Similarly, a virtual circuit is used exclusively for the duration of a single data transaction, but may be allocated to other users after the transaction is complete.

The network flags are used to indicate the type of data carried; user information or system management. They also carry information about the network congestion. One byte of error checking and correcting code is added. Only the header is checked and corrected, the integrity of the data carried is the responsibility of the end user.

22.4.1 The ATM switch

Multi-media LANs use intelligent 'hubs' for switching. A hub is the centre of a star topology network into which many nodes are fed. ATM networks depend on switching for connectivity. ATM standards specify switch interfaces, not the internal design of the switches, so there is a choice of ATM switch architecture and implementation. Two ATM switch architectures are: (1) bus-based and (2) parallel, which has distinct advantages.

In a bus-based architecture, switch ports are connected to a shared backplane bus. Cells enter the switch then wait until the bus becomes available before they can be routed to the destination port. This is a form of time division multiplexing. Bus performance will always be limited by printed circuit boards and connectors. Backplane clocking speeds are limited by the laws of physics.

A parallel architecture does not share a data bus. Switch ports connect to a switch matrix and cells are routed to their destination port by switching elements. Arrays of switching elements can be used to build larger switches. Data does not have to wait since paths through the switch are in parallel (space division multiplexing).

Both ATM switch architectures are shown in *Figure 22.5*.

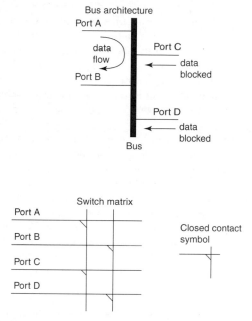

Figure 22.5 ATM switch architecture

22.4.2 *ATM transmission*

High speed data transmission systems such as SDH or SONET will be used for transporting the ATM cells. Any system can be used to carry ATM cells, but those with T1 or E1 rates are likely to be the lowest commonly used.

22.5 Transmission control protocol/Internet protocol (TCP/IP)

Since 1983, TCP/IP packets have been used to send data across local area networks (LANs) and wide area networks (WANs), in addition to the Internet. The underlying network can be either; ethernet, token ring, fibre distributed data interchange (FDDI) or WAN technology. Details of TCP/IP and the Internet are given in Chapter 23.

23 The Internet

The Internet is global and the number of users is growing rapidly. It supports electronic commerce and has the ability for businesses to reach a global market. Security and electronic payment techniques are being developed, which are encouraging the introduction of a wide range of commercial applications and services. Search engines enable users to find information on a vast range of subjects. The Internet supports education and training using on-line interactive programs.

The Internet was designed as a data network, but can also support audio and video traffic. New network protocols are being developed that will make efficient use of the network and give guaranteed quality of service (QoS). Widespread use of audio and video applications on the Internet would require significantly increased bandwidth in the Internet backbone and increased access bandwidths. Internet technologies are also used to advantage in corporate networks, or Intranets. The capabilities of the Internet have grown steadily, with key developments including the introduction of electronic mail, and the file transfer protocol (FTP) in the 1970s, the domain name system and TCP (a reliable transport protocol) in the early 1980s.

In 1989 the World Wide Web (WWW) was proposed, which became the key application of the 1990s by transforming the Internet into a global, multimedia information service. The WWW attracted a wide range of users and led to the explosive growth to give an estimated 150 million Internet users by March 2001. What started as a network for communication and data sharing by academics and government researchers has become widely used from the home and from the office, to access a vast pool of information published on the WWW. There has been a gradual evolution of on-line shopping, audio and video applications (including telephony and video-conferencing).

23.1 Internet control

There are several bodies that oversee the development of technical standards on which the Internet is based. The Internet Society provides long-term support for Internet standards developed by the

Internet Engineering Task Force (IETF). The Internet Architecture Board (previously called the Internet Activities Board) associates its activities with the Internet Society. The IAB oversees the architecture for the protocols used by the Internet and the processes used to create them. The management of the process for Internet standards development are handled by the Internet Engineering Steering Group (IESG) under the auspices of the Internet Society.

The work of developing and standardizing the Internet protocol suite is undertaken by the IETF. The IETF has been expanding the capabilities of the Internet protocol (IP) suite, particularly those concerned with real-time traffic and quality of service. The IETF is also developing new intelligent networking and new services such as multicast and mobile IP. Action is being taken to enable the Internet to support the full ISO 10646 character-set that contains characters for most languages.

Another 'standards' organization is the World Wide Web Consortium (W3C). The work of the W3C includes hypertext mark-up language (HTML), the integration of object technology into the Web, and improvements to HTTP (hypertext transfer protocol). The Joint Electronic Payment Initiative (JEPI) was initiated by W3C to standardize methods of payment between buyers and sellers on the WWW.

23.2 Internet terminology

The evolution of the Internet has depended on a number of different technologies, each playing a specific role.

23.2.1 Internet protocol

The foundation of the Internet is the Internet protocol (IP), which routes data packets across the network from one computer to another. IP is very simple because it focuses specifically on packet routing but does not guarantee packet delivery. Packets may be lost or corrupted, and a sequence of packets sent between two hosts may take different routes and may arrive in a different order. These issues are left for other protocols to handle.

IP packets are sent from one router to another towards their destination. Each router maintains routing tables, which determine the output interface on which a packet should be sent. When a packet arrives at a router, it is stored in a queue. When it reaches the front of the queue, the router reads the destination address in the header. This address is looked up in the routing table and then the router transmits the packet out of the appropriate interface port. Routing tables are regularly updated to take account of network failures or changes to the configuration of the network.

If the network is congested, the input queues at routers can become very long, which increases the packet transit delay. Under heavy congestion the queues may become full and packets are discarded. It is up to other protocols (such as TCP) to recognize a missing packet and to request retransmission.

The military origin of IP has helped it to become the *de facto* internetworking standard. The lack of delivery-guarantee makes it simple and efficient, because not all applications require these guarantees. Also, its ability to adapt to network failures and configuration changes has made IP particularly robust. Its ability to adapt has enabled it to cope with the exponential Internet growth and allows networks of many sizes and type to be easily connected.

23.2.2 *IPv4*

Currently, all IP networks use Version 4 of the Internet protocol. This was developed before it was realized how popular IP would become. The greatest limitation of IP version 4 (IPv4) is the size of the address space, which allows only a 32-bit address. This is not enough for everyone to be allocated a unique address. Hence IP addresses have become scarce and schemes have had to be developed to enable sufficient addresses. One technique is network address translation (NAT), which translates and isolates the Internet address so that previously used addresses can be re-used on the internal network side of the router.

The Internet Engineering Task Force (IETF) has been working to design the next generation of IP, IPv6 (the proposed IPv5 was rejected). A global experimental network has been built, called the 6bone, with the aim of evaluating early IPv6 implementations and identifying any problems. The IPv4 header contains 10 fields, a checksum, two addresses and some options, see *Figure 23.1*. IPv4 addresses are based on a 32-bit word that, in theory, gives IPv4 over 4 billion addresses. The options field in the IP header is a disadvantage because it requires recalculation of a checksum for every possible IPv4 header type.

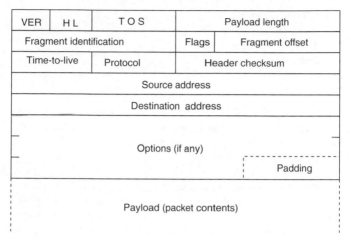

Figure 23.1 IPv4 header

IP header fields

VER (version) = 4 bits

The Version field indicates the format of the header. In this case it is version 4.

HL (header length) = 4 bits

The header length is the length of the IP header in 32 bit words and points to the beginning of the payload. Note that the minimum value for a correct IPv4 header is 5, but this value is increased if the options field is used. The maximum header length is 60 octets (including options).

TOS (type of service) = 8 bits

The type of service provides an indication of the quality of service desired. These parameters are used when transmitting a packet through a particular network and are a guide to the selection of the actual service. Some networks treat high precedence traffic as more important than other traffic. During periods of high load, only traffic above a certain level of precedence is accepted. The choice is a trade-off between low-delay, high-reliability and high-throughput.

Bits 0–2 determine the precedence, given by the following values:

111 – Network Control
110 – Inter-network Control

101 – CRITIC/ECP
100 – Flash Override
011 – Flash
010 – Immediate
001 – Priority
000 – Routine

Bit 3 determines the delay (0 = normal delay, 1 = low delay). Bit 4 determines the throughput (0 = normal, 1 = high). Bit 5 determines the reliability (0 = normal, 1 = high) and Bits 6–7 are reserved for future use.

Total length = 16 bits

Total length is the length of the packet, measured in octets, including the header and payload. This field allows the length of a packet to be up to 65 535 octets. Such long packets are impractical for most hosts and networks; Ethernet based networks limit packets to 1500 octets.

All hosts must be prepared to accept packets of up to 576 octets (whether they arrive whole or in fragments). It is recommended that hosts only send packets larger than 576 octets if they have checked that the destination is prepared to accept them. The number 576 is selected to allow a reasonable sized data block to be transmitted in addition to the required header information. For example, this size allows a data block of 512 octets plus 64 header octets to fit in a packet. The maximal IP header is 60 octets, but is typically 20 octets, thus allowing a margin for headers of higher level protocols.

Fragment identification = 16 bits

An identifying value assigned by the sender to aid in assembling the fragments of a packet.

Flags = 3 bits

Various control flags. Bit 0 is reserved and must have a value of zero. Bit 1 (DF) (0 = fragmentation allowed, 1 = do not fragment). Bit 2 (MF) (0 = last fragment, 1 = more fragments).

Fragment offset = 13 bits

This field indicates where in the packet this fragment belongs. The fragment offset is measured in units of 8 octets (64 bits). The first fragment has offset zero.

Time-to-live = 8 bits

This field indicates the maximum time the packet is allowed to remain within the Internet. If this field contains the value zero, then the packet must be destroyed. The value in this field is reduced by at least one during IP header processing by a router. The 'time-to-live' is measured in units of seconds, but if the router processes the packet in less than one second the T-T-L will be shorter than expected. In fact, the T-T-L must be thought of as an upper limit on the time a packet may exist. The aim is to discard undeliverable packets.

Protocol = 8 bits

This field indicates the next higher level protocol used in the payload portion of the IP packet. The values for various protocols are specified in 'assigned numbers', and examples are: 06 (hex) = TCP, 11 (hex) = UDP and 58 (hex) = IGRP (Cisco proprietary).

Header checksum = 16 bits

This is a checksum on the header only. Since some header fields change (e.g. time to live), this is computed and verified at each point that the IP header is processed.

The header is broken down into 16-bit words, which are then summed using one's complement arithmetic with any carry added in. A one's complement of the result gives the checksum. For the purpose of this computation, zero value is used for the checksum field. This checksum is simple to compute and has been found to be adequate.

Source address = 32 bits

The source address.

Destination address = 32 bits

The destination address.

Options: variable length

The options field is used mainly for IP packet tracing, time-stamping and security. In some cases the security option may be required in all packets. The option field is variable in length; there may be zero or more options.

There are two formats for options. The first is a single octet of option-type. The second format is a sub-packet containing an option-type octet, an option-length octet and option data of variable length. The option-length given in the second octet is the length of the whole sub-packet.

The option-type octet is viewed as having 3 fields: 1 bit is the copied flag, 2 bits are the option class and 5 bits are the option number. The copied flag indicates that this option is copied into all fragments on fragmentation (0 = not copied, 1 = copied).

The option classes that are used are 0 for control and 2 for debugging and measurement. Option classes 1 and 3 are reserved for future use.

The following IP options are defined:

Class	Number	Length	Description
0	0	–	End of option list. This option occupies only one octet; it has no length octet.
0	1	–	No operation. This option occupies only one octet; it has no length octet.
0	2	11	Security. Used to carry security, restrictions compatible with DOD requirements.
0	3	variable	Loose Source routing. Used to route the IP packet based on host supplied information.
0	9	variable	Strict Source routing. Used to route the IP packet based on host supplied information.
0	7	variable	Record route. Used to trace the route an IP packet takes.
0	8	4	Stream ID. Used to carry the stream identifier.
2	4	variable	Internet timestamp.

Specific option definitions

Padding: between 1 and 3 octets long

The IP header padding is used to ensure that the IP header ends on a 32-bit boundary. The value of the padding is zero.

23.2.3 IPv6

The IPv6 header has six fields and two addresses, see *Figure 23.2*. The address space in the IP header is 128 bits and gives the possibility of 3.4×10^{38} addresses. In IPv6 the options field is placed after the IPv6 header and before the data. Thus the IPv6 header has a fixed size and hence a checksum field is not needed. This also makes IPv6 headers suitable for processing by hardware, owing to their fixed size.

The IPv6 header has the following fields:

1 Version, a 4-bit field that identifies the IP version used.

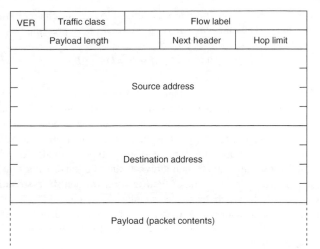

VER	Traffic class	Flow label		
Payload length			Next header	Hop limit

Source address

Destination address

Payload (packet contents)

Figure 23.2 IPv6 header

2 Traffic class, an 8-bit field that is used for distinguishing between different classes or priorities of IPv6 packets. These bits may be used to offer various forms of differentiated service (i.e. priority) for IP packets.

3 Flow label, a 20-bit field that is used to identify a flow of IP packets between a particular source and destination.

4 Payload length, a 16-bit field that gives the length of the IPv6 payload.

5 Next header, an 8-bit field that identifies the type of header immediately following the IPv6 header, for example, UDP or TCP. This is the same as used in IPv4.

6 Hop limit, an 8-bit field whose decimal value is decreased by one each time the packet passes through a router. The IP packet is discarded if this value reaches zero.

7 Source address, a 128-bit field that represents the source address of the IPv6 packet.

8 Destination address, a 128-bit field that represents the destination address of the IPv6 packet.

Each IPv6 address is represented by eight groups of 16-bit numbers, separated by colons. Each number is displayed as four hexadecimal figures. An example of an IPv6 address is

$$16EE : 3B40 : 0000 : 0000 : 000C : 03E9 : 7ADE : 9BB7.$$

Where there are zeroes in a hexadecimal block, a double-colon can replace them. Thus the same IPv6 address can be represented as:

$$16EE : 3B40 :: C : 3E9 : 7ADE : 9BB7.$$

IPv6 has been designed to support 'real-time' traffic, such as voice and video. It does this using the traffic class field and the flow label field, and these are used to minimize the delay and delay variation of the IP packets. The traffic class field is assigned different values (1 to 8) dependent upon the type of IP data, typically video could be assigned a low value and e-mail a high value, the lower values giving greater priority. The flow label field is used to identify a stream of IPv6 packets that each have a particular destination address and a particular source address. A router will be able to identify packets from their flow label and simply forward them without having to read the address fields.

It is probable that an IPv6 Internet will be operated in parallel with the existing IPv4 Internet. Host computers will contain both the IPv4 and IPv6 stack, and applications using the Internet will determine the IP protocol version from information provided by the domain name service (DNS).

23.2.4 Transport protocols

Transport protocols are used together with IP in order to provide error checking, error correction or recovery. The transport frame (including data and transport header) is placed inside the Internet packet and forms the data part of an IP packet.

There are two main transport protocols in use on the Internet – UDP and TCP. User packet protocol (UPP) is intended for sending messages without guarantee of arrival and without notifying the sender of successful or failed delivery. It is very simple. In addition to the data carried within the packet, information is supplied about the application being used and a checksum is transmitted to indicate whether any data has been corrupted in transit. In contrast to UDP, the TCP protocol is designed to handle all types of network failure. If packets are lost or corrupted then TCP arranges for them to be re-transmitted. If packets arrive out of order then TCP will re-order them. If packets are repeatedly lost, the TCP source will assume that the network is congested and reduce its transmission rate.

23.2.5 Ports

At any one time, a terminal connected to an IP network may be interacting with other hosts and a number of different applications may be involved. Port numbers are used to identify for which application and for which activity a packet is intended. Each UDP or TCP packet carries two port numbers; one port number identifies the server application and the other port number is selected by the client to distinguish the particular activity. Commonly used ports, which are used with applications like Telnet and FTP, are defined in the Assigned Numbers RFC. An RFC is a request for comments and is part of the standardization process used by the IETF.

23.2.6 Telnet

Telnet is a very basic Internet function that allows a user to access a server remotely. This is known as remote terminal access and works by carrying ASCII text (typed by the user) to the remote server and then returning the output from the application on the remote server to the user.

23.2.7 File transfer protocol (FTP)

FTP provides a basic service for the reliable transfer of files from one machine to another and allows the user to establish a control connection between their client and the server. This connection can be used to navigate through the server's directory structure and request the transmission of files. A separate data connection is set up to transmit the files.

23.2.8 Domain name service (DNS)

The domain name service provides name/address translation for all objects on the Internet. Every computer (and every router) on the Internet has an individual name written by concatenating two or more names using '.' as a separator, e.g. ieee.org. In this name the top level domain is 'org' and the IEEE owns the sub-domain (or name-space) 'ieee'.

The owner of a 'name space' must run a DNS server, which presents its address to the DNS server at the next level up. Applications such as FTP, SMTP, Telnet and WWW send a request to the local DNS server, which responds by producing the answer itself or with the address of a DNS server that can provide the answer. If the local DNS server responds with the answer, then this may have come either from its own look-up tables or from another DNS server (that it approached on the application's behalf).

23.2.9 E-mail

Electronic mail or e-mail is the electronic equivalent of the traditional postal service. E-mail is sent from a mail 'client' (a program which runs on the user's machine) to the destination mailbox using the simple mail transfer protocol (SMTP). Other protocols such as POP3 or IMAP4 are used for checking and retrieving mail from a mailbox. SMTP routes e-mails from the mail agent to the destination mailbox via a number of mail handlers. Mail handlers behave like conventional sorting offices; that is, they sort the mail and pass it on to the next mail handler.

An SMTP mail address is based on an Internet domain name and takes the form user@domain, e.g. steve.winder@ieee.org. In order to route the message to an e-mail address of this form, the mail agent depends on DNS to translate the domain part of the e-mail address (the part after '@') into an IP address.

An e-mail can be sent to a single recipient or a list of people with equal ease. E-mail lists are a simple way of informing a group of people. Unfortunately, e-mail lists can distribute junk mail with equal ease and can be used to spread computer viruses (as attachments).

23.2.10 Bulletin boards

An alternative to e-mail lists are bulletin boards and newsgroups. These are the electronic equivalent of a 'paper and pins' notice-board. They allow users to post messages for viewing by user groups. Anyone can reply to the original message.

Both e-mail lists and bulletin boards may be used to support group working, but there is an important difference. With e-mail, messages are sent to each user on a mailing list, so all the recipients receive the latest information automatically. However, with bulletin boards the user must choose to visit the site to read new postings.

23.2.11 World Wide Web

The application that has caused the growth in Internet use is the World Wide Web (WWW). The WWW makes publishing, display and retrieval of all kinds of data very easy and now accounts for most of the traffic on the Internet. Documents on the WWW are usually written in hypertext mark-up language (HTML). HTML is a program listing that describes the layout of the page and the components on it, such as the text and images. HTML pages are stored at servers and the information is accessed using a client application known as a browser; the most popular browsers are Netscape Navigator and Microsoft's Internet Explorer.

A Web page is identified by a Uniform Resource Locator (URL). The browser and web-server usually communicate using hyper-text transfer protocol (HTTP), but HTTP is not the only protocol supported by browsers (for example, FTP and Telnet are also supported). Therefore, the URL must specify the file to be retrieved and the protocol to be used. In general a URL takes the form:

⟨protocol⟩ : //⟨hostname⟩ : ⟨port⟩⟨directory⟩⟨filename⟩

WWW pages can use a wide variety of data types including text, image, animation, audio or video. The web-server specifies the type of data using a multipurpose internet mail extension (MIME), which was originally designed for sending multimedia e-mail. The browser can handle some MIME content types, for example text/plain, text/HTML or image/GIF. Other formats such as application/postscript, audio/MPEG or video/MPEG are normally handled by a separate application, known as a 'helper' or a 'plug-in'.

The components (such as text, pictures, sound, etc.) of an HTML-based document may be distributed across a number of servers. Only the HTML program/framework has to exist on the web-server. The program includes both a URL and fetch instruction for each component required on the page, for processing by the WWW client. The content presented at a WWW client may be from stored files or from dynamic generation when the program is run on the server. The common gateway interface (CGI) is used to communicate to the web-server and is commonly used to perform searches.

23.3 Internet structure

The Internet has grown by connecting a number of existing networks. This has resulted in a mesh of networks, each owned by a different organization. When a packet of data travels across the Internet it will enter and leave via Internet access providers and may pass through any number of 'IP backbone' providers en-route. Internet access providers may also be backbone providers. Customers are connected to their Internet access provider's point of presence (POP) through dial-up lines (PSTN or ISDN) or leased lines. Backbone providers are connected to other backbone providers either via peer agreements (where they agree to carry each other's traffic) or via an exchange such as LINX or CIX.

Some access providers are re-sellers that have bought connections from a single backbone provider. These commercial agreements can result in packets being routed via indirect routes, such as when different Internet access providers within the same country route to each other via the US and thereby increase congestion. However, commercial pressure is leading to a more hierarchical structure using national Internet exchanges and the emergence of global backbone providers.

23.4 Intranets and firewalls

Internet technologies are also used for internal company networks, known as an intranet. An intranet is an internal company network based on the Internet protocol (IP) and makes use of WWW server and client technology, e-mail, etc. An intranet may also include links that cross the public Internet to connect together a company's different local area networks. The key benefits and opportunities of an Intranet include better communications, internal e-mail, mailing lists, etc., and allows more effective publishing and distribution of information within an organization (employee handbooks, quality management systems, etc.) which need version control.

Connection between the Intranet and the Internet has security risks, including access by hackers or competitors to commercial information. Encryption can be used to make e-mail secure and restricting access to information can be achieved by installing a 'firewall' between the internal network and the Internet.

Firewalls are installed at the network interface (usually on a router) and are either packet screens or proxy servers. Packet screens examine each packet passing to and from the internal network. Packets are either allowed through or discarded depending on the rules applied to the firewall. Proxy servers control the type of services that may pass through the firewall, for example WWW or e-mail. Proxy servers are very generally secure.

23.4.1 Virtual private networks (VPNs)

Confidence in the security of IP networks is increasing and intranets are now being provided on a shared IP infrastructure, on what is referred to as a virtual private network (VPN). Potentially, thousands of VPNs can be provided over a single shared high-capacity global IP network. VPNs can also support Extranets, in which secure connectivity is extended to suppliers, customers or communities of interest over a common IP infrastructure.

23.5 Electronic commerce

Many companies now have on-line shops through which goods can be ordered. Travel, banking and insurance companies have been particularly active, but security is an issue. Cryptography provides the security necessary for privacy of communications and payments to be made over the Internet. Two forms of encryption are in common use: symmetric encryption and asymmetric encryption (also known as public-key cryptography). Both use cryptographic algorithms and keys to encode and decode data. The keys are parameters used in the mathematical encryption process.

In symmetric key systems, the sender and receiver must use identical keys. The security of symmetric key systems depends on keeping the key information private, therefore the key must be sent over a separate and secure path. This is a disadvantage for widespread use in electronic commerce. Symmetric encryption algorithms are much faster than public-key algorithms and the digital encryption standard (DES) is widely used for this.

Public-key schemes use pairs of related keys. Each user has a private key, which is kept secret, and a public key, which is published and readily available to others. Encryption of a message with the recipient's public key provides confidentiality because the owner of the corresponding private key is the only person who can read it. This approach is often used as a secure way of exchanging symmetric keys. Encryption of a message with the sender's private key also provides a signature since the message can only be decrypted correctly using the sender's public key.

The overall security of key-based systems depends on the strength of the cryptographic algorithms and on the security of the keys. The strength of the encryption increases exponentially with key length.

23.6 Network computing

In network computing all software and data is stored remotely on servers in the network and downloaded to the user's PC as and when required. The main advantages of network computing are reduced version management, and rapid access to new applications and services.

23.6.1 *Java*

The basis of network computing is the Java programming language. The software required to execute Java is available within every WWW browser, which means that almost every networked computer is capable of running a Java program. Java offers animation and the full power of a computer programming language in the browser. Java applications can expect to be seen running independently of the browser.

23.7 Real-time services

Limits in processing speed and Internet bandwidth have restricted the development of real-time services. But recent developments in PC technology have enabled video compression within a realistic period. Competition in telecommunications has created considerable Internet bandwidth. The high speed processors and wide bandwidth have led to a rapid increase in the speed of real-time services development over the Internet, including Internet telephones, audio and video streaming applications.

The prospect of 'free' calls using Internet telephony has excited some enthusiasts, but the truth is the fundamental costs of Internet telephony and PSTN telephony are similar. However Internet telephony has yet to match the convenience, ease of use, call quality and customer service of the PSTN. However, Internet telephony is easily integrated with other computer applications and so is likely to continue to be used and to develop. A gateway is needed to provide PSTN interconnectivity for telephone calls between IP networks and the PSTN.

Voice over IP is offered on a 'best effort' basis and the quality of the call is dependent upon network congestion. Factors that affect call quality are packet loss and wide variations in network delay. The IETF is developing the integrated services architecture (ISA), which is a framework that aims to offer control over the quality of service provided by a network. ISA allows applications to prioritize their traffic and to request network resources to enable the priority system to work. Two major components of ISA are IP multicast and resource reservation protocol (RSVP).

The quality of service seen by the user depends on the quality of service provided by the network, the operating systems and the application protocols. The requirement for a reliable signal conflicts with the requirement for low delay. Retransmission of a packet found to contain errors will lead to increased end-to-end delay and increased jitter (variation in end-to-end delay).

The enhancement of the Internet and intranets to support real-time services is exciting, because it promises a future in which a single network can be used for voice, data and all other media. It also promises excellent computer/telephony integration (CTI) and interactive entertainment.

23.7.1 Differentiated services

One approach to providing quality of service (QoS) in core IP networks is based on the IETF standard of differentiated services (Diffserv). Diffserv gives high priority to certain types of data; for example, gold service for delay-sensitive voice or video traffic, silver for medium-priority services such as e-mail, and bronze for low-priority data. Priorities can be identified in several ways, including the type of service bits in the IP packet header. One technique for giving different levels of priority is weighted random early discard (WRED), which selectively discards low priority packets at the edge of the network to protect the core from congestion. Another technique is class-based queuing (CBQ), in which bandwidth and delay limits are set. Packets are then processed through the router according to their class or priority.

23.8 Wireless application protocol (WAP)

Wireless application protocol (WAP) provides the equivalent of HTML on mobile terminals. Originally it was intended to provide a translation of wireless mark-up language (WML), but this failed and hence WAP-enabled pages have to be published on the Internet to enable mobile terminals to view them.

24 Organizations

Telecommunications have evolved from each country having its own public network operators (PTOs). When telecommunications consisted mainly of telephone and teleprinter facilities most operations were on a monopoly basis, and in most countries the monopoly was a government-sponsored organization, frequently combined with the postal services.

The use of standards in telecommunications has become more important now that different network operators need to interface, because of the international nature of telecommunications and because of deregulation. One value in having a standard is that manufacturers can produce identical components in greater quantities, thus reducing the overhead costs of each part. The main value of having a standard is to remove the chaos that could arise if different network operators were allowed to use their own specifications. One has only to travel across Europe to see all the different types of telephone socket (about 50 types are in use). Standards organizations have been set up by countries, or groups of countries such as the European Community (EC).

Standards can also be linked to network regulators, such as OFTEL in the UK or the FCC in the USA. These organizations are like the police in that they authorize the activities of network operators. They make sure that the network operators and telecommunication equipment manufacturers apply the standards bodies' recommendations. They also intervene to stop unfair trading practices, to allow free competition between companies.

24.1 Network regulators

In the UK, the Telecommunications Act 1981 effectively removed the monopoly on the supply of terminal equipment, achieving deregulation and allowing suppliers to offer suitable telephones, telex, facsimile, cordless telephones, answering machines, etc., direct to customers for connection to the public network. The PTO became 'British Telecom' but retained its monopoly status as a network operator and its right to assess and approve equipments before they could be connected to the public network. The Telecommunications Act 1984 removed the monopoly status of BT to operate public networks, and BT became

British Telecom plc, a public company, able to operate and compete on a commercial basis.

The issue of licences to operate public networks is the responsibility of the Department of Trade and Industry (DTI) on the advice of the Office of Telecommunications (Oftel), an independent body set up to regulate the telecommunications industry. In addition to the licences issued to BT, and other public network operators; licences are also issued for the operation of local cable television and value-added networks (VANs) accessed via the public networks.

Other EC countries are having to introduce deregulation to comply with the 1992 free competition requirement, sometimes against opposition from the monopoly operators. This does not necessarily imply privatization of a nationalized network operator, but should result in freedom for other operators to compete.

Deregulation in the USA began in earnest in 1984 when the giant AT&T was forced to split the Bell network, creating 'Baby Bells'. Seven Regional Bell Operating Companies (RBOCs) were formed. Each was allocated an area – not defined from geographic features, such as state borders, but defined from the telecommunication system infrastructure. *Figure 24.1* shows the approximate division of the USA, with regions 1 through to 7 marked.

Figure 24.1 RBOC areas

The regions are operated by the RBOCs given in *Table 24.1.*

Long distance carriers such as AT&T, MCI Communications and Sprint have been competing in the long distance market for several years, but the RBOCs have had a virtual monopoly in the local and regional market. To force competition in this area, the US Government has introduced The Telecommunications Act of 1996. This law was designed to allow competition in the USA's local and long distance telecommunications market. It also has sections to regulate television broadcast information, including regulations to reduce the risk of sexually explicit material reaching minors.

Table 24.1 Regional allocations
for 'Baby Bells'

Region	RBOC
1	Pacific Telesis
2	Q WEST
3	Ameritech Services
4	South Western Bell
5	Bell South Services
6	VERIZON
7	Nynex

It is the role of the Federal Communications Commission (FCC) to administer the 1996 act. Section 251 refers to interconnection between telecommunications carriers. Each carrier is obliged to: interconnect with all other carriers; provide number portability; free access to telephone numbers, etc. In addition, the charges each carrier can apply to others for the use of cable ducts, poles and other rights of way must be reasonable. Obstructive behaviour, such as having incompatible features, is not allowed, nor are restrictions on the resale of services.

24.2 The International Telecommunications Union (ITU)

The success of a national or international communications network rests on the ability to obtain a satisfactory link between calling and receiving terminals, via a number of tandem connected links; hence the need to specify the maximum permissible distortion in each individual transmission link, and its interface characteristics. Ideally, interface specifications should be such that interfaces are compatible, requiring no additional interface units, or at least specified sufficiently rigidly to enable satisfactory interface units to be provided. The performance specification for each individual link must be such that the longest international call should still provide acceptable quality.

The need for standardization was realized as far back as 1932 when the International Telecommunications Union (ITU) was set up. The problem has become much more acute since the Second World War, and a number of organizations are now contributing to the standardization process. A vast amount of documentation now exists on interface and network recommendations and specifications, and in the following chapters reference is made to various definitive documents containing detailed information.

The ITU is an international union between member states, from an amalgamation of the International Telegraph Union and the International Radiotelegraph Union. In 1949 ITU became a specialized agency of the United Nations, and has its headquarters in Geneva, Switzerland.

24.2.1 Branches of the ITU

The International Radio Consultative Committee (CCIR), the International Telegraph and Telephone Consultative Committee (CCITT) have always been part of the ITU. To make this clear, they have now been renamed as ITU-R and ITU-T respectively.

The ITU-R and ITU-T hold a Plenary Assembly every 4 years to draw up a list of relevant technical subjects to be studied in the following 4 years by their study groups, and to consider the recommendations made by the study groups for the previous 4 years. If adopted the recommendations are published in the ITU books, a new set of volumes appearing every 4 years.

The ITU-T and ITU-R issue recommendations, which are not mandatory specifications. The recommendations have to consider the established interests of national operators, which can involve alternative recommendations (e.g. the 30 channel TDM system now adopted in Europe and the 24 channel adopted in America, Canada and Japan). The recommendations are widely quoted in purchasing specifications and in the interface specifications for terminal equipment connected to the public network. They also form the basis for most of the specifications issued by other standardization bodies. The extensive use made of ITU-T recommendations illustrates the importance of the role it plays in the standardization and specification process for telecommunications. The relevant ITU-T document numbers are referred to in the following chapter.

24.3 CEN/CENELEC and CEPT

CEN (Comité Européen de Normalisation) and CENELEC (Comité Européen de Normalisation Electrotechnique) are European standards organizations with their secretariat in Brussels and with members drawn from the European standards organizations. Their main activity is in the field of information technology, and they are actively engaged in the preparation of functional standards for Open System Interconnect (OSI). CENELEC covers the electrotechnical aspects with the International Technical Commission (IEC), and CEN covers other aspects in conjunction with International Organization for Standardization (ISO).

CEN/CENELEC issue European standards documents known as ENVs in the draft stage, and ENs when finalized. The ENV is essentially a European experimental standard which is used for a trial period of 2 to 3 years before being adopted as a definitive EN specification. The documents providing the background of the functional standards programme are:

M-IT-01: The concept and structure of Functional Standards
M-IT-02: Directory of Functional Standards

The European Conference of Postal and Telecommunications Administrations (CEPT) was established in 1959 to foster closer cooperation between the European operators. It was non-political and represented the interests of the network operators. It is therefore closely associated with the work of ITU and CEN/CENELEC who share the common aim – the agreement to common European standards which permit satisfactory operation between all European public network operators.

CEN/CENELEC and CEPT have set up a coordinating body, the Information Technology Steering Committee (ITSTC) to coordinate the standardization activities where they overlap. CEPT submits recommendations to the Technical Recommendations Advisory Committee (TRAC) who select those recommendations which could be converted into European telecommunications standards. These documents are known as NETs, and have similar status to the ENs issued by CEN/CENELEC. NETs are based closely on ITU-T recommendations, but in order to create definitive European standards, the alternatives allowed in ITU-T recommendations are deleted and a single standard for adoption by all EC and EFTA countries is presented.

Most of the NETs issued, or in course of preparation (NETs 1 to 10), are 'ACCESS' NETs dealing with the characteristics of terminal equipments at a network interface. A list of NETs is given in Chapter 25.

24.4 European Telecommunications Standards Institute (ETSI)

ETSI, which has its headquarters in Sophia Antipolis in southern France, evolved from CEPT following proposals made in an EC green paper in 1987. Unlike CEPT and CEN/CENELEC, ETSI has as its members representatives of manufacturers, users, private service providers, national standards and research bodies in addition to the network operators. It also has representatives of the EC and EFTA who act as councillors.

The task of ETSI is to produce standards for approval by the ETSI technical assembly. On approval these will be known as Interim European Telecommunications Standards (I-ETS). After a trial period these will be voted upon by the member countries, and the I-ETS will become ETS. As with the NETs issued by CEPT, the ETSs aim to eliminate alternatives, and the majority vote on the I-ETS provides the moral justification to impose the standards on all EEC and EFTA members. It is possible that Article 90 of the Treaty of Rome may be invoked as a means of imposing change on authorities who resist enforced standardization.

A further objective of enforced standards is to achieve a free European market. National specifications are successfully employed as a means of restricting imports, and this has been combined by some European nations with national approval procedures which insist on a national approval certificate before a product can be imported. EC action is therefore being taken not only to introduce mandatory specifications for telecommunications equipment to apply throughout the EC, but also to insist that each member country accepts the equipment approvals of other EC countries.

24.5 American National Standards Institute (ANSI)

ANSI has administered and coordinated a voluntary standards system since 1918. Although it was founded by US government agencies and engineering societies, it remains privately run. ANSI represents the US interests in ISO and, indirectly, the IEC. ANSI was a founding member of ISO.

Committees within the ANSI umbrella develop standards for use in the USA. These are then taken forward to ISO or IEC standards meetings with the hope of becoming worldwide standards. Although ANSI does not itself examine products to see if they meet the required standards, it does provide the necessary accreditation of these third parties. The institute is working to gain worldwide acceptance of product certifications from these third parties.

24.6 International Organization for Standardization (ISO)

ISO is a worldwide federation of standards bodies from about 100 countries. These include ANSI from the USA, DIN from Germany and BSI from the UK. One of the most familiar standards are the ISO photographic film speed codes (ISO 100, ISO 400, etc.). Another is the DIN derived paper sizes (ISO 216 Standard) of A0, A1, through to A5 which are in common use. In the telecommunications field, the Open Systems Interconnection (OSI) 7-layer model is the best known ISO standard. Quality standards such as the ISO 9000 series are also well known.

Strangely, ISO does not mean International Standards Organization. ISO is a word derived from the Greek 'isos', which means 'equal'. ISO works in English, French and Russian, so an acronym could not be used and understood in all three languages.

24.6.1 International Electrotechnical Commission (IEC)

Standards in electrical and electronic engineering are the responsibility of the IEC which works with ISO. This organization was founded in 1906 as a result of a resolution made at the International Electrical Congress, held in St. Louis, USA in 1904. There are members from 54 countries participating in the work of the commission. One of the best known standards is that of the electrical plug used to carry mains power, the IEC320.

24.7 Electrical Industry Association (EIA) and Telecommunications Industry Association (TIA)

The EIA has been providing equipment standardization since 1924. The modem interface RS232C has been developed, but instead of becoming RS232-D it is now known as the EIA232-D standard. The EIA represents the electrical industry, raising issues of concern to its members and keeping its members informed of developments likely to affect them.

The United States Telecommunications Suppliers Association (USTSA) and the Information and Telecommunications Technologies Group of the EIA (EIA/ITG) merged to form the TIA in 1988.

The TIA is accredited by ANSI and contributes towards developing voluntary standards which promote trade.

24.8 Conformance testing and certification organizations

The principle that product validation by any one EC or EFTA country should be accepted by all other partners (memorandum M-IT-03) is leading to an overhaul of the organization of national standards and approval procedures. Each country has built up its own specification, approval and quality procedures, which must now be harmonized if the EC open market conditions are to be enforced properly.

The UK source of national standards is the British Standards Institute (BSI). Although specifications for telecommunications equipment may still be issued by BSI, they must now conform with any relevant EC specifications (ENs and NETs) if they are to be enforceable.

Conformance testing, previously carried out by the network operators themselves, or by a national organization, must now be carried out by an independent organization authorized by the EC to carry out this work. To achieve this the CEC (Commission of European Communities) has a Conformance Testing Services (CTS) programme to approve and appoint the organizations in the member countries authorized to supervise conformance testing.

The EC scheme is known as the European IT (Information Technology) certification scheme, and will result in the issue of European IT certificates for approved equipment. At network level the scheme will be administered by the National IT Certification Coordinating Member (a government or delegated national agency), and each such national body will be represented on the European Committee for IT Certification (ECITC). In the UK this body is BABT (British Approvals Board for Telecommunications). The National IT Certification Coordinating Member will normally delegate testing and certification to organizations which they have approved for particular classes of equipment (for which they have the necessary test facilities).

Equipment for approval has first to be submitted to an approved test house, who test to specification and pass the results back to the manufacturer. If satisfactory, those results can then be submitted to the approval certification body, authorized to issue the European IT Certificate.

Accredited test houses are assessed against ISO guides 25, 38, 43 and 45, and the technical requirements laid down by the Testing

Support Service (TSS), for the types of equipment that they wish to test. Satisfactory equipment will display a CE mark to show that it has met all the necessary standards. In the USA, ANSI accredits organizations such as the TIA and Underwriters Laboratories Inc. (UL) to act as test houses.

24.9 Organization addresses

Organization	Address
ANSI	American National Standards Institute 11 West 42nd Street, 13th Floor New York, NY 10036 USA Tel. (212) 642 4900 Fax. (212) 398 0023 Internet: http://www.ansi.org
AMTA	American Mobile Telecommunications Association 1150 18th Street NW Suite 250 Washington, DC 20036 USA Tel. (202) 331 7773 Fax. (202) 331 9062
Bellcore	Bell Communications Research Customer Service 60 New England Avenue Piscataway, NJ 08854 USA Tel. (800) 521 2673
CSA	Canadian Standards Association 178 Rexdale Blvd Canada, M9W 1R3 Tel. (416) 747 4000
CTIA	Cellular Telecommunications Industry Association 1250 Connecticut Avenue NW Suite 800 Washington, DC 20036 USA Tel. (202) 785 0081

EIA	Electronics Industries Association 2001 Pennsylvania Ave. N.W. Washington, DC 20006 USA Tel. (202) 457 4966
ETSI	European Telecommunications Standards Institute 06921 Sophia-Antipolis Cedex France Tel. +33 92 94 4200 Internet. http://www.etsi.fr
FCC	Federal Communications Commission 1919 M Street NW Washington, DC 20554 USA Tel. (202) 418 0200 Fax. (202) 418 0232
GED	Global Engineering Documents 1990 M Street W Suite 4000 Washington, DC 20036 USA Tel. (800) 854 7179 Tel. (202) 429 2860 Rapidoc(R) Willoughby road Bracknell Berkshire RG12 8DW UK Tel. (01344) 861666 Fax. (01344) 714440
IEC	International Electrotechnical Commission 3, Rue de Varembre PO Box 131 1211 Geneve 20, Switzerland Tel. +41 22 919 0211 Fax. +41 22 919 0300

The IEE
Michael Faraday House
Six Hills Way
Stevenage
Herts, SG1 2AY
UK
Tel. 01438 767290
Fax. 01438 742856
www.iee.org.uk

ComSoc
IEEE Communications Society
305 East 47th Street
New York, NY 10017
USA
Tel. (212) 705 8900
Fax. (212) 705 8999

IEEE
Institute of Electrical and Electronics
Engineers
Three Park Avenue 17th Floor
New York, NY 10016 5997
USA
Tel. (212) 419 7900
Fax. (212) 752 4929
www.ieee.org

ISO
International Organization for
Standardization
1, Rue de Varembre
PO Box 56
1211
Geneve 20, Switzerland
Tel. +41 22 34 12 40

ITU (formerly CCITT)
International Telecommunications
Union
Place de Nations
1211
Geneve 20
Switzerland

NCTA
National Cable Television Association
1724 Massachusetts Avenue NW
Washington, DC 20036
USA
Tel. (202) 775 3669

NECA	National Exchange Carrier Association
	80 South Jefferson Road
	Whippany NJ 07981
	USA
	Tel. (973) 884 8207
	Fax. (973) 884 8469
NIST	National Institute of Standards and
	Technology
	Technology Building 225
	Gaithersburg, MD 20889
	USA
NRTC	National Rural Telecommunications
	Co-operative
	212 Co-operative Way
	Herndon VA 20171
	USA
	Tel. (703) 787 0874
	Fax. (703) 464 5300
NTCA	National Telephone Co-operative
	Association
	4121 Wilson Boulevard 10th Floor
	Arlington, VA 22203
	USA
	Tel. (703) 351 2000
	Fax. (703) 351 2001
PCIA	Personal Communications Industry
	Association
	500 Montgomery Street
	Suite 700
	Alexandria, VA 22314
	USA
	Tel. (703) 739 0300
	Fax. (703) 836 1608
TIA	Telecommunications Industry
	Association
	2500 Wilson Boulevard Suite 300
	Arlington
	VA 22201-3834
	USA
	Tel. (703) 907 7700
	Fax. (703) 907 7727

UL	Underwriters Laboratories, Inc.
	333 Pfingsten Road
	Northbrook, IL 60062
	USA
	Tel. (800) 676 9473
	Fax. (708) 272 8129

USTA	United States Telecom Association
	1401 H Street NW Suite 600
	Washington, DC 20005 2164
	USA
	Tel. (202) 326 7300
	Fax. (202) 326 7333

WCA	Wireless Communications Association
	International
	1140 Connecticut Avenue Suite 810
	Washington, DC 20036
	USA
	Tel. (202) 452 7823
	Fax. (202) 452 0041

25 Standards

Legally enforced standards include safety, Electromagnetic Compatibility (EMC), Electromagnetic Interference (EMI) and hazard warnings. Unless these standards are met it would be virtually impossible to sell or use equipment.

Voluntary standards make the item marketable, such as approved connector types. Quality standards, like the ISO 9000 series, come into this category. A label saying that the manufacturer has met the ISO 9001 quality standard indicates that the product design has been checked at all stages. Also, it indicates that full documentation exists and the product will continue to meet the same performance. Meeting this standard also shows that the finished product was inspected and tested using calibrated test equipment. Other quality standards in this series are ISO 9002 which covers production, installation and testing. ISO 9003 covers field inspection and testing by a supplier.

Other standards, such as the ISO seven-layer reference model, are built into system specifications. An example of this is the use of the seven-layer model in ATM and Frame Relay system designs.

25.1 Interface and performance standards

Interfaces are created wherever dissimilar equipments are connected. The first and most obvious interfaces occur where terminal equipment is connected to the public network. The public network consists of a large number of communication links, some analogue and some digital, with various channel contents, and using cable, optical fibre or radio links as the transmitting media. These all have to be capable of interconnection and create a further series of interfaces. The growth of international communications where the equipment used by various countries may differ, and the introduction of satellite communications, have also led to interface problems. Finally the changeover from mechanical to digital switching (sometimes with an intermediate stage of electronic switching) requires interfacing between the different forms of switching in use.

In order to operate correctly each interface must have both electric and procedural compatibility. Electrically the signals must be compatible in formation, frequency or repetition rate, level, etc., and so must the interface impedances and mode of operation (duplex, simplex, semi-duplex, asynchronous, synchronous, etc.).

The procedures by which calls are set up, monitored and supervised are referred to as protocols. These have to cover the signalling arrangements for routing, answering, charging and clearing, and in the highly flexible systems made possible with the stored program control (SPC), also include various forms of validation, message repeat signals, call holding, etc.

The rapid advances in terminal equipment, particularly that for business use, have resulted in a steady increase in the complexity and number of alternative interfaces. The original telex, for instance, was a simple single-speed system, whereas the more recent teletext employs modems capable of providing alternative bit speeds and modulating methods over the PSTN. These have to be matched to the receiving station equipment before transmission can commence.

Whenever interfaces are incompatible, equipment is required to convert from one interface to the other. This equipment must achieve electrical, signal, and protocol compatibility, and usually involve modems or codecs (coder, decoders). In the case of FDM or TDM signals, the interface equipment may require restoration of these signals to the original baseband signal to achieve a proper interface match.

25.2 Analogue telephone network standards in Europe

Standardization within the European Union for all analogue telephone networks will allow a telephone in say, Germany, to be unplugged and connected into the Spanish telephone network. At present telephone systems use different connectors. The working voltage and current levels on analogue lines are country dependent.

The UK standard is BS 6305. This and other European standards have been incorporated into a NET 4 standard (some 1000 pages long!). Another standard, NTR 3 has replaced NET 4, and this uses common tests whilst allowing for different national networks. The aim is to replace all existing standards by a single unifying standard called CTR 21.

25.3 USA telecommunications approval: FCC part 68

There are few performance requirements laid down by the FCC rules (part 68). The majority of requirements are to protect the network and to prevent interference to other users.

Mains power fed equipment must have good isolation between the mains supply and the telephone line connections. A standard pair of tests is to apply 1 kV rms AC between one telephone line connection point and an earth bonding point, and then repeat the test for the other telephone line connection point. In both tests, a current of no more than 10 mA should flow. A second pair of tests is to apply 1.5 kV between each of the mains power supply terminals and an earth bonding point. The current flow should measure less than 10 mA.

Signal power levels are to be less than −9 dBm in total across the whole of the voice band (200 Hz to 4 kHz). However, DTMF tones used for dialling and signalling can have a level of 0 dBm (this is the combined power of both tones). These power levels are measured in a 600 Ω load. At frequencies above the voice band, the permissible power levels are lower than −14 dBV (note the change of units to dBV, because line impedances are lower than 600 Ω above the voice band). The maximum signal power level reduces with frequency; above 100 kHz signal power levels are limited to −55 dBV.

Other FCC part 68 rules include: the ringing load (Ringing Equivalence Number, or REN); the off-hook AC impedance of 600 Ω; the nominal off-hook DC resistance of 200 Ω; etc.

25.4 Open systems interconnection (OSI)

The creation of operating standards interrelates with the establishment of the Open Systems Interconnection (OSI), under the auspices of the International Standards Organization (ISO) working in cooperation with other standards bodies, computer manufacturers and user groups. The main objective of OSI is to enable dissimilar and incompatible equipment to exchange data and computer information by means of a set of agreed protocols. This covers only part of the problem outlined in Section 25.1, being concerned primarily with digital data. New digital systems are now being designed to satisfy OSI requirements; in particular ISDN (Integrated Services Digital Network) satisfies OSI and permits simultaneous digitized voice transmissions.

The overall framework for OSI is contained in the ISO publication IS7498 – Information Processing Systems – Open Systems Interconnection – Basic Reference Model.

OSI is based on an abstract reference model which defines the functional requirements enabling two computer systems to communicate over the public networks. It is designed to provide the greatest possible degree of freedom for the internal workings of a

314

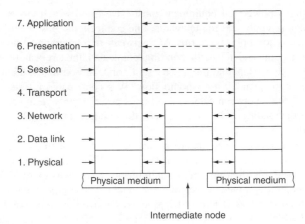

| 7. Application → |
| 6. Presentation → |
| 5. Session → |
| 4. Transport → |
| 3. Network → |
| 2. Data link → |
| 1. Physical → |

Physical medium   Physical medium

Intermediate node

Figure 25.1   Structure of OSI reference model

computer, but to define its external behaviour to enable both sending
and receiving computers, which may be of different manufacture, to
communicate using the public network. This puts some constraints on
both the computer terminals and the public network which must be
capable of interfacing using the OSI model.

The structure of the reference model is shown in *Figure 25.1*.
There are seven layers, each layer representing a key function. Each
layer is responsible for providing a service to the layer above it, and
for maintaining a relationship with the equivalent layer in the receiving terminal with which it is communicating. Each layer requires two
types of standard, the first representing in abstract terms the functions
contained within the layer, and service provided to the layer above,
and the second the protocol used within the layer to communicate
between systems. The functions of the layers are as follows:

Physical layer: handles the electrical and mechanical interface
to the public network, including the procedures for activating and
deactivating the call and establishing the compatibility of the signals
between terminals and network.

Date link layer: provides synchronization and error control for the
transmitted data.

Network layer: provides addressing and routing information to
establish, maintain and terminate the connection between the terminals.

Transport layer: is the lowest layer in which relay is not allowed,
and is the lowest level in which the protocol always operates between

the terminals, as shown in *Figure 25.1*. The layer provides terminal-to-terminal control and information interchange.

Session layer: is the lowest layer dealing explicitly with end-to-end communications, whereas layers below are involved in the interconnections process. The layer controls the session services to be provided (e.g. duplex, semi-duplex, simplex, etc.).

Presentation layer: evaluates the translation processes required to make the information formats of the two terminals compatible.

Application layer: provides an interface between the communications environment and the end-user application process.

Standards relating to OSI are issued by ISO and ITU-T as listed in *Table 25.1*.

Table 25.1 Standards relating to OSI

	ISO	CCITT
Reference Model	IS 7498	X200
Service Definition Conventions	DP 8509	X210
Network Services	IS 8348	X213
Transport Service	IS 8072	X214
Session Service	IS 8326	X215
Transport Protocol	IS 8073	X224
Session Protocol	IS 8327	X225
Link Layer Access Protocol	IS 7776	X25
Link Layer Multilink Protocol	DIS 7478	X25
Network Layer Protocol	IS 8208	X25

IS: International Standard
DIS: Draft International Standard
DP: Draft Proposal

A similar concept to OSI is ONP (Open Network Provision) and ONA (Open Network Architecture), originating in the United States supported by the FCC (Federal Communications Commission).

25.5 Conformance testing and certification

The principle that product validation by any one EC or EFTA country should be accepted by all other partners (memorandum M-IT-03) is leading to an overhaul of the organization of national standards and approval procedures. Each country has built up its own specification, approval and quality procedures, which must now be harmonized if the EC open market conditions are to be enforced properly.

The UK source of national standards is the British Standards Institute (BSI). Although specifications for telecommunications equipment may still be issued by BSI, they must now conform with any relevant EC specifications (ENs and NETs) if they are to be enforceable.

Conformance testing, previously carried out by the network operators themselves, or by a national organization, must now be carried out by an independent organization authorized by the EC to carry out this work. To achieve this the CEC (Commission of European Communities) has a Conformance Testing Services (CTS) programme to approve and appoint the organizations in the member countries authorized to supervise conformance testing.

The EC scheme is known as the European IT (Information Technology) certification scheme, and will result in the issue of European IT certificates for approved equipment. At network level the scheme will be administered by the National IT Certification Coordinating Member (a government or delegated national agency), and each such national body will be represented on the European Committee for IT Certification (ECITC). In the UK this body is BABT (British Approvals Board for Telecommunications). The National IT Certification Coordinating Member will normally delegate testing and certification to organizations which they have approved for particular classes of equipment (for which they have the necessary test facilities).

Equipment for approval has first to be submitted to an approved test house, who test to specification and pass the results back to the manufacturer. If satisfactory, those results can then be submitted to the approval certification body, authorized to issue the European IT Certificate. All products sold in Europe must also be CE marked to show that they conform to EC regulations.

Accredited test houses are assessed against ISO guides 25, 38, 43 and 45, and the technical requirements laid down by the Testing Support Service (TSS), for the types of equipment that they wish to test.

In the USA, the FCC determines the standards for connection to the telephone network.

25.6 Electromagnetic compatibility (EMC) and electromagnetic interference (EMI)

In addition to the performance specifications to which equipment must be tested to obtain IT certificates, all equipment must now satisfy strict EMC requirements. These requirements cover both the limits of interference that the equipment can produce and the levels of outside interference to which it must be immune.

The harmonized standards set by CENELEC are EN50082 – 1 for domestic or light industrial equipment. EN55022 applies to IT equipment. All equipment must be tested for EMC as part of the approval procedure before being issued with the European IT Certificate. This testing requires specialized test equipment and may be carried out either by a separate test house or the same test house if they have the necessary equipment.

From the design and manufacturing aspects, reduction of both types of EMC require careful attention to power lead filtering, screening of sensitive circuitry, and disposition of earth returns.

The FCC part 15 regulations used in the USA have two levels of emission limits. The Class A limit applies to equipment for industrial use. The Class B limit applies to equipment intended for residential use. The limits are frequency dependent. For example, using the Class A limit (measured at 10 m from the equipment); radiated emissions of 39 dμMV are permitted in the frequency range 30–88 MHz, but in the 88–216 MHz range the limit is 43.5 dBμV.

Electrostatic discharge and susceptibility to external fields are covered by another standard. This is met by IEC 801 (parts 2–4), now incorporated into the European standards EN 61000-4-2 through to EN 61000-4-4.

25.7 The Low Voltage Directive

The EC Low Voltage Directive (LVD) has been mandatory since 1975. However, there was no evidence required to show that this requirement had been met. On January 1, 1997 the EC made its compliance a requirement of CE marking. A Declaration of Conformity must be issued with the product, giving details of model number, year of manufacture, factory location, dealer information, etc. The manufacturer must maintain a technical file with the evidence to back up this information. The manufacturer may be called upon to produce this file during an ISO 9001 inspection, or as evidence in case of legal action after an accident or incident involving the product.

The LVD covers all equipment designed for use with a voltage rating of 50 V to 1 kV AC, and 75 V to 1.5 kV DC. Not a standard itself, the directive indicates which standards must be met. In terms of telecommunications, any equipment operating at these voltages must comply with EN 60950 (IEC 950).

25.8 ISO 9000

The acceptance of equipment under the EC approvals scheme must carry with it safeguards to ensure that the quality and performance of the product is sustained for the duration of the approval. This involves a further procedure to monitor the manufacturing, test facilities, and documentation of the manufacturer. Such assessments have been in existence for many years, and were originally carried out by large organizations on their suppliers. The BS 5750 approvals scheme was designed to provide an approval system that would be generally acceptable to all customers and remove the need for separate assessments. The need to harmonize with other EC and EFTA countries, and link in with the European IT certification, has resulted in overall responsibility being transferred to the ISO, and BS 5750 being superseded by ISO 9000.

Manufacturing approval is normally based on a 'Quality Manual', which the manufacturer must produce and submit for acceptance. The quality manual must detail the complete manufacturing procedures from the content of parts lists and drawings, ordering, goods inward inspection, through manufacture with details of special processes, component handling, stage inspection, to test, packing and despatch. At all stages considerable stress is placed on records and segregation and on traceability, the means by which any part of the equipment may be traced back to pinpoint a problem area. Approved equipment should be assembled from approved components, and the documentation has to ensure not only that the approved components are ordered, but also that 'Certificates of Conformity' for the components are supplied and filed for future reference. The quality manual also has to deal with subjects such as operator training, the lines of responsibility within the company, and the checks carried out by company management to ensure that the procedures are being strictly adhered to.

In the UK the responsibility for assessing and approving manufacturers under ISO 9000 schemes is vested in the BSI, who now provide a service, paid by the company seeking approval, to approve the quality system and check that the required procedures and paperwork are installed in accordance with the quality manual. Periodic visits are then required to check that no deviations from the manual are occurring, and that the procedures laid down in the manual are effective in achieving the desired quality level in the company's products. The products to be assessed in this manner may be hardware or software, or a combination of both.

25.9 Measurement standards

The equipment and manufacturing approval schemes depend for their success on the proper maintenance of measurement standards, with the accuracy of the measuring equipment used in design, manufacture and maintenance regularly checked against higher standards, and recalibrated where necessary. The regular calibration of test equipment by an authorized independent calibration centre is an essential part of a manufacturer's quality plan, and such equipment usually carries a label stating the date of the most recent calibration.

An authorized calibration centre has substandards against which calibration is carried out, and which have to be traceable back to the national standards.

25.10 Reliability and failure statistics

With the strict standardization and quality controls described above it is possible to produce statistical data on component failure from which the probability of faults on components and complete equipment can be predicted. The data also provides a basis for acceptance testing with random sampling.

25.10.1 Mean time between failures

The BSI and IEC definition of reliability is 'the characteristic of an item expressed by the probability that it will perform a required function under stated conditions for a stated period of time'. It is important to understand that there is a fundamental connection between reliability and probability; the most probable time for which a component, or a module, or indeed a system will work without failure is referred to as the mean time between failures (MTBF). This is a statistically derived conception and represents the average time between failures of a large number of samples.

In an electronic assembly it is usual to assess the MTBF from a knowledge – obtained by experience, or from the manufacturer – of the failure rate of all the component parts, and then multiply these by the number of parts. The failure rate is the reciprocal of MTBF. See, for example, *Table 25.2* for a simple subassembly of 10 transistors, 30 resistors, 2 capacitors and 2 connectors.

320

Table 25.2

	Failure rate % per 1000 hours*	No.	Product
Transistors	0.0025	10	0.025
Resistors	0.001	30	0.03
Capacitors	0.01	2	0.02
Connectors	0.1	2	0.2
Total		44	0.275

* These figures are as an example only, and do not necessarily reflect true values

The failure rate for this subassembly is 0.275% per 1000 hours. Thus

$$\text{MTBF} = \frac{1}{\text{failure rate}}$$

$$= \frac{1}{0.275} \times 1000 \times 100$$

$$= 363\,636 \text{ hours or approx. 40 years}$$

Most components exhibit a high failure in their early life, followed by a long period of almost constant low failure rate, which then finally increases again as the component enters the 'wear-out' period preceding its end of life. This is the so-called bath-tub curve (*Figure 25.2*).

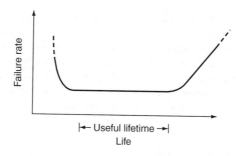

Figure 25.2

The failures associated with the initial part of this curve are practically eliminated by 'burning in' components – and equipment – for a period during which the items concerned are subjected to enhanced environmental stress. For example, elevated temperature and/or humidity or temperature cycling between, say, 0°C and 85°C.

Quality may be defined as 'fitness for intended purpose' and current national and international specifications for electronic components permit the purchase of components of 'assessed quality'. The UK BS 9000 series of specifications, and European CECC 401 201, for example, appertain to metal film resistors; and supplies to these specifications can be certified with traceable authority from the manufacturer to comply in all respects with the specifications.

25.10.2 Quality and acceptance testing

Acceptance testing for parts of assessed quality involves taking a truly random sample from every batch manufactured and subjecting it to all the measurements demanded by the published specification. The size of the sample chosen depends on the chosen value of acceptable quality level (AQL); this is the maximum average percentage of defective items in the sample which is considered tolerable. The sample size may also depend on historical results of previous samples; that is, if consistently marginal values of AQL are found over a period, the sample size could be increased for the same AQL. Statistical tables are published by BSI (ISO 9001) and MOD (DEF 131A) which give sample size in relation to batch size for a selected AQL and level of inspection.

25.10.3 Fault probability and distribution

Various mathematical formulae have been proposed to show the relation between a measured value and its frequency of occurrence among a randomly selected number of samples of the item involved. Where the number of observations or measurements is large, the distribution is known as Gaussian or 'normal' and is symmetrical about a centre line representing the most frequently occurring measurement.

The area under this curve (*Figure 25.3*) lying between the limits of the 'standard deviation' σ is 68%, that is 68% of measurements may deviate from the centre value (node) by amounts equal to or less than $\pm\sigma$. Similarly 95% of measurements lie between $\pm2\sigma$ and 99.7% of measurements lie between $\pm3\sigma$

If we take a y axis through the mean value $\bar{x}$ and put $\omega = (x - \bar{x})/\sigma$, then

$$y = \frac{1}{\sigma\sqrt{2\pi}} e^{-\omega^2/2}$$

Normal distribution is usually associated with component sampling; if we take n_s samples from a batch n_B, and let μ_s be the sample mean and σ_s the sample standard variation, then mean of batch is

$$\mu_s \pm \frac{\gamma \sigma_s}{\sqrt{n_s}}$$

where

$\gamma = 1$ for a confidence level of 68%
$\gamma = 2$ for a confidence level of 95%
$\gamma = 3$ for a confidence level of 99.7%

Let us consider an example.

In a batch sample of 100 resistors of nominal value 10 k, the standard deviation is 100 ohms. The standard error of the mean is $100/\sqrt{100} = 10$ ohms or 0.01 k. Hence with 95% confidence we can state that the values of the lot lie between $(10 \pm 2 \times 0.01)$ k, that is between 10.02 k and 9.98 k.

Other forms of probability distribution (e.g. Poisson, exponential and Weibull) are also used in reliability calculations. The Poisson distribution is applicable where events occur at random; it postulates the probability of an event p occurring m successive times in n observations as

$$p(m) = \frac{(np)^m \, e^{-np}}{m!}$$

The Poisson distribution mean value is np and the standard deviation in $\sqrt{np}$.

An exponential distribution is defined as

$$y = \frac{1}{x} e^{-x/\bar{x}}$$

where $\bar{x}$ is the mean value. Here 37% of the lot will be above and 63% below the mean value. As an example, if it is assumed that failures of the equipment vary exponentially, and the MTBF is 1000 hours, what is the probability it will work for 700 hours or more without failure? Here

$$\frac{x}{\bar{x}} = \frac{700}{1000} = 0.7$$

and the area under the curve beyond $x = 0.7$ is 0.497. Hence there is a 49.7% probability that the equipment is still working after 700 hours.

The Weibull distribution is given by

$$y = \alpha\beta(x - \gamma)^{\beta-1} \exp[-\alpha/(x - \gamma)\beta]$$

where α is a scale factor, β is a shape factor (note that if $\beta = 1$, y becomes exponential) and γ is a location factor dependent on the position of the y axis.

25.10.4 *Equipment availability*

Another concept applied to repairable equipment is that of availability (a):

$$a = \frac{\text{MTBF}}{\text{MTBF} + \text{MTTR}} \quad \text{where MTTR is Mean Time To Repair}$$

Clearly for serially connected modules $1, 2, 3, \ldots, n$ the overall availability is the product $a_1 \times a_2 \times a_3 \times \cdots \times a_n$. However, availability may be improved by paralleling identical modules, in which the availability then becomes

$$1 - (1 - a_1)(1 - a_2)$$

for two units in parallel. This is known as parallel redundancy and the trade-off is of course the doubling of the equipment cost and increase in its size. However, if, for example, the single-module availability is 60%, then parallel redundancy improves this to $1 - (0.4 \times 0.4)$, that is 84%.

25.11 Summary of ITU-T recommendation documents

25.11.1 *Serial data transmission over telephone circuits*

V1	Equivalence of binary notation and the two conditions (1,0) code
V2	Power levels for data transmission over telephone lines
V3	International Alphabet No.5
V4	General structure of signals of International Alphabet 5 for data transmission over public networks
V5	Modulation and data signalling rates for synchronous data transmission on switched networks
V6	As V5 but on leased telephone circuits
V8	Modem call set-up handshaking protocol
V8*bis*	Identification and selection of common operating modes between data circuit-terminating equipments (DCEs) and between data terminal equipments (DTEs) over PSTN or leased lines

V10	Electrical characteristics of unbalanced double-current interchange circuits for data communications circuits using ICs
V11	As V10 but for balanced circuits
V13	Answerback unit simulator
V14	Transmission of start-stop characters over synchronous bearer channels
V15	Data transmissions using acoustic coupling
V16	Transmission of analogue medical data using modems
V17	A 2-wire modem for facsimile applications with rates up to 14 400 bit/s
V18	Operational and interworking requirements for DCEs operating in the text telephone mode
V19	Modems for parallel data transmission using telephone signalling frequencies
V20	Modems for parallel data transmissions on general switched networks
V21	300 baud modem for switched telephone network
V22	1200 bit/s full duplex two-wire modem for switched telephone network
V22*bis*	2400 bit/s full duplex two-wire modem for switched telephone network
V23	600/1200 bit/s modem for switched telephone network
V24	List of definitions for interchange circuits between data terminal equipment and the associated modems
V25	Automatic calling/answering equipment for general switched telephone networks
V26	2400 bit/s for four-wire point-to-point circuits
V26*bis*	2400/1200 bit/s modems for switched telephone networks
V27	4800 bit/s modems for leased circuits
V27*bis*	4800/2400 bit/s modems with adaptive equalizer for leased circuits
V27*ter*	4800/2400 bit/s modem for switched telephone network
V28	Electrical characteristics for unbalanced double-current interchange circuits
V29	9600 bit/s modem for leased circuits

V31	Electrical characteristics for single-current interchange circuits controlled by current closure
V32	New standard for full duplex modems for high-speed operation PSTN or leased lines
V33	A 4-wire modem operating at 14 400 bit/s over point-to-point leased circuits
V34	28.8 kbps modem for the PSTN
V34 (1996)	This version of the V34 standard provides for 33.6 kbps PSTN modems
V35	48 kbit/s data transmission using the 60–108 kHz FDM group band
V36	Modems for synchronous data transmissions using the 60–108 kHz group band
V40	Error indication with electromechanical equipment
V41	Code-independent error control system
V42	Error-correcting procedures for DCEs using asynchronous-to-synchronous conversion
V42*bis*	Data compression procedures for data circuit-terminating equipment (DCE) using error correction procedures
V43	Data flow control
V50	Limits for transmission quality of data transmission
V51	Maintenance of international-telephone-type circuits used for data transmission
V52	Distortion and error-rate-measuring apparatus for data transmission
V53	Maintenance of telephone-type circuits used for data transmission
V54	Loop test devices for modems
V55	Impulsive noise-measuring instruments for telephone-type circuits
V56	Comparative tests for modems on telephone-type circuits
V57	Data test set for high-speed data signalling
V58	Management information model for V-Series DCEs
V61	Analogue simultaneous voice and data transmission
V70	Digital simultaneous voice and data
V75	DSVD terminal control procedures
V75 Append. II	Session establishment using V.75/H.245 procedures

V76	Generic multiplexer using V.42 LAPM-based procedures
V80	Videophone
V90	56 kbps modem
V92	A digital modem and analogue modem pair for use on the PSTN at data signalling rates of up to 56 000 bit/s downstream and up to 44 000 bit/s upstream
V100	Interconnection between public data networks (PDNs) and the PSTN
V110	Support of data terminal equipments with V-series type interfaces by an integrated services digital network (This Recommendation is also included but not published in I series under alias number I.463)
V120	Support by an ISDN of data terminal equipment with V-series type interfaces with provision for statistical multiplexing (This Recommendation is also included but not published in I series under alias number I.465)
V130	ISDN terminal adaptor framework
V140	Procedures for establishing communication between two multi-protocol audio-visual terminals using digital channels at N × 64 kbit/s or N × 56 kbit/s
V230	General data communications interface layer 1 specification
V250	Serial asynchronous automatic dialling and control (previously known as V.25 ter)
V251	Procedure for DTE-controlled call negotiation (previously known as Annex A, V.25 ter)
V252	Procedure for control of V.70 and H.324 terminals by a DTE
V253	Control of voice-related functions in a DCE by an asynchronous DTE
T4	Group 3 facsimile protocol

25.11.2 Data networks

X1	International user classes of service in public data networks
X2	International user facilities in public data networks

X3	Packet assembly/disassembly (PAD) facility in a public data network
X4	Signal structure of International Alphabet No. 5 for data transmission over data networks
X5	Facsimile Packet Assembly/Disassembly facility (FPAD) in a public data network
X6	Multicast service definition
X7	Technical characteristics of data transmission services
X8	Multi-aspect PAD (MAP) framework and service definition
X20	Interface between data terminal and data circuit terminating equipment (DTE and DCE) for start/stop transmission on public data networks
X20*bis*V21	Interface between DTE and DCE for synchronous operation on public data networks
X21*bis*	Interface between DTE on public data network and synchronous V series modems
X22	Multiplex DTE/DCE interface for user classes 3–6
X30	Support of X21, X21*bis* and X20*bis* based DTEs by an Integrated Services Digital Network (ISDN) [This Recommendation is also included but not published in I series under alias number I.461]
X31	Support of packet mode terminal equipment by an ISDN [This Recommendation is also included but not published in I series under alias number I.462]
X32	Interface between DTE and DCE for terminals operating in the packet mode and accessing a Packet-Switched Public Data Network through a public switched telephone network or an Integrated Services Digital Network or a Circuit-Switched Public Data Network
X33	Access to packet-switched data transmission services via frame relay
X34	Access to packet-switched data transmission services via B-ISDN
X35	Interface between a PSPDN and a private PSDN which is based on X25 procedures

X36	Interface between DTE and DCE for public data networks providing frame relay
X36 Amend. 1	Switched virtual circuit (SVC) signalling and refinements of permanent virtual circuit (PVC) signalling
X36 Amend. 2	Frame transfer priority
X36 Amend. 3	Frame discard priority, service classes, NSAP signalling and protocol encapsulation
X37	Encapsulation in X25 packets of various protocols including frame relay
X38	G3 facsimile equipment/DCE interface for G3 facsimile equipment accessing the Facsimile Packet Assembly/Disassembly facility (FPAD) in a public data network situated in the same country
X39	Procedures for the exchange of control information and user data between a Facsimile Packet Assembly/Disassembly (FPAD) facility and a packet mode Data Terminal Equipment (DTE) or another FPAD
X42	Procedures and methods for accessing a public data network from a DTE operating under control of a generalized polling protocol
X45	Interface between data terminal equipment (DTE) and data circuit-terminating equipment (DCE) for terminals operating in the packet mode and connected to public data networks, designed for efficiency at higher speeds
X46	Access to FRDTS via B-ISDN
X48	Procedures for the provision of a basic multicast service for data terminal equipments (DTEs) using Recommendation X25
X49	Procedures for the provision of an extended multicast service for data terminal equipments (DTEs) using Recommendation X25
X51	Parameters for the international interface between multiplex synchronous data networks using a 10 bit envelope structure
X51*bis*	Parameters for the international interface between 48 kbit/s synchronous data networks using a 10 bit envelope structure

X52	Encoding of anisochronous signals into a synchronous user bearer
X53	Number of channels on international multiplex links at 64 kbit/s
X54	Allocation of channels on international multiplex links at 64 kbit/s
X55	Interface between synchronous data networks using a 6 + 2 envelope structure and single channel per carrier (SCPC) satellite channels
X56	Interface between synchronous data networks using an 8 + 2 envelope structure and single channel per carrier (SCPC) satellite channels
X57	Method of transmitting a single lower speed data channel on a 64 kbit/s data stream
X58	Fundamental parameters of a multiplexing scheme for the international interface between synchronous non-switched data networks using no envelope structure
X60	Common-channel signalling for data circuit switching systems
X61	Data user part of Signalling System No. 7
X70	Terminal and transit control signalling system for start-stop services on international circuits between anisochronous data networks
X71	Decentralized terminal and transit control signalling system on international circuits between synchronous data networks
X75	Terminal and transit call procedures and data transfer systems on international packet switched data networks
X76	Network-to-network interface between public data networks providing frame relay
X76 Amend. 1	Switched virtual circuits
X76 Amend. 2	Frame relay service classes and priorities
X77	Interworking between PSPDNs via B-ISDN
X80	Signalling systems for circuit switched data services
X81	Interworking between an ISDN circuit-switched and a circuit-switched public data network (CSPDN)

25.11.3 *Data communications networks – OSI (Open Systems Interconnection)*

X227	Connection-oriented protocol for the association control service element in OSI: Protocol specification
X227*bis*	Connection-mode protocol for the Application Service Object Association Control Service Element in OSI
X228	Reliable transfer: Protocol specification
X229	Remote operations: Protocol specification
X233	Protocol for providing the OSI connectionless-mode network service: Protocol specification
X234	Information technology – Protocol for providing the OSI connectionless-mode transport service
X234 Amend. 1	Addition of connectionless-mode multicast capability
X235	OSI Connectionless Session protocol: Protocol specification
X236	OSI Connectionless presentation protocol: Protocol specification
X237	Connectionless protocol for the OSI association control service element: Protocol specification
X237 Amend. 1	Incorporation of extensibility markers and authentication parameters
X237*bis*	Connectionless protocol for the OSI Application Service Object Association Control Service Element
X244	Procedure for the exchange of protocol identification during virtual call establishment on packet switched public data networks
X245	OSI Connection-oriented Session protocol: Protocol Implementation Conformance Statement (PICS) proforma
X246	OSI Connection-oriented presentation protocol: Protocol Implementation Conformance Statement (PICS) proforma
X247	Protocol specification for the OSI Association Control Service Element: Protocol Implementation Conformance Statement (PICS) proforma
X248	OSI Reliable transfer: Protocol Implementation Conformance Statement (PICS) proforma

334

336

X328	General arrangements for interworking between public data networks providing frame relay data transmission services and Integrated Services Digital Networks (ISDNs) for the provision of data transmission services
X340	General arrangements for interworking between a Packet-Switched Public Data Network (PSPDN) and the international telex network
X350	General interworking requirements to be met for data transmission in international public mobile satellite systems
X351	Special requirements to be met for Packet Assembly/Disassembly facilities (PADs) located at or in association with coast earth stations in the public mobile satellite service
X352	Interworking between Packet-Switched Public Data Networks and public maritime mobile satellite data transmission systems
X353	Routing principles for interconnecting public maritime mobile satellite data transmission systems with public data networks
X361	Connection of VSAT systems with packet-switched public data networks based on X25 procedures
X400	Message handling service for text communications and electronic mail
X402	Information technology – Message Handling Systems (MHS): Overall architecture
X408	Message handling systems: Encoded information type conversion rules
X411	Information technology – Message Handling Systems (MHS): Message transfer system: Abstract service definition and procedures
X413	Information technology – Message Handling Systems (MHS): Message store: Abstract service definition
X419	Information technology – Message handling systems (MHS): Protocol specifications
X420	Information technology – Message handling systems: Interpersonal messaging system
X421	Message handling systems: COMFAX use of MHS

X622	Protocol for providing the connectionless-mode network service: Provision of the underlying service by an X25 sub-network
X623	Protocol for providing the connectionless-mode network service: Provision of the underlying service by a sub-network that provides the OSI data link service
X625	Protocol for providing the connectionless-mode network service: Provision of the underlying service by ISDN circuit-switched B-channels
X630	Efficient Open Systems Interconnection (OSI) operations
X633	OSI Network Fast Byte Protocol
X634	OSI Transport Fast Byte Protocol
X637	Basic connection-oriented common upper layer requirements
X638	Minimal OSI facilities to support basic communications applications
X639	Basic connection-oriented requirements for ROSE-based profiles
X641	Quality of service: framework
X642	Quality of service – Guide to methods and mechanisms
X650	OSI Basic reference model: Naming and addressing
X660	Procedures for the operation of OSI Registration Authorities: General procedures
X662	Procedures for the operation of OSI Registration Authorities: Registration of values of RH-name-tree components for joint ISO and ITU-T use
X665	Procedures for the operation of OSI Registration Authorities: Application processes and application entities
X666	Procedures for the operation of OSI Registration Authorities: Assignment of international names for use in specific contexts
X669	Procedures for the operation of OSI Registration Authorities: Registration procedures for the ITU-T subordinate arcs

X670	Procedures for registration agents operating on behalf of organizations to register organization names subordinate to country names
X671	Procedures for a registration authority operating on behalf of countries to register organization names subordinate to country names
X680	Abstract Syntax Notation One (ASN.1): Specification of basic notation
X681	Abstract Syntax Notation One (ASN.1): Information object specification
X682	Abstract Syntax Notation One (ASN.1): Constraint specification
X683	Abstract Syntax Notation One (ASN.1): Parameterization of ASN.1 specifications
X690	ASN.1 encoding rules: Specification of Basic Encoding Rules (BER), Canonical Encoding Rules (CER) and Distinguished Encoding Rules (DER)
X691	ASN.1 encoding rules – Specification of Packed Encoding Rules (PER)
X700	Management framework for OSI for ITU-T applications
X701	OSI Systems management overview
X702	OSI Application context for systems management with transaction processing
X703	Open Distributed Management Architecture
X703 Amend. 1	Support using Common Object Request Broker Architecture (CORBA)
X710	OSI Common Management Information Service
X711	OSI Common Management Information Protocol: Specification
X712	OSI Common Management Information Protocol: Protocol Implementation Conformance Statement (PICS) proforma
X720	OSI Structure of management information: Management information model
X721	OSI Structure of management information: Definition of management information
X722	OSI Structure of management information: Guidelines for the definition of managed objects
X723	OSI Structure of management information: Generic management information

X920	Open distributed processing – Interface definition language
X930	Open distributed processing – Interface references and binding
X950	Open distributed processing – Trading Function: Specification
X952	Open distributed processing – Trading Function: Provision of trading function using OSI directory service

25.11.4 ISDN (Integrated Services Digital Network)

I110	Structure of I series recommendations
I111	Relation with other recommendations relevant to ISDN
I112	Vocabulary of terms for ISDN
I113	Vocabulary of terms for broadband aspects of ISDN
I114	Vocabulary of terms for universal personal telecommunication
I120	ISDN – Description
I121	Broadband aspects of ISDN
I122	Framework for frame mode bearer services
I130	Services supported by, and network capabilities of, ISDN
I140	Attribute technique for the characterization of telecommunication services supported by ISDN and network capabilities of ISDN
I141	ISDN network charging capabilities attributes
I150	B-ISDN ATM functional characteristics
I200	Guidance to the I.200-Series of Recommendations
I210	Principles of telecommunications services supported by ISDN
I211	Bearer services supported by ISDN (e.g. 64 kbit/s)
I212	Teleservices supported by ISDN (e.g. telephony, teletext, videotext)
I220	Common dynamic description of basic telecommunication services
I221	Common specific characteristics of services
I230	Definition of bearer service categories
I231.1	Circuit-mode 64 kbit/s unrestricted, 8 kHz structured bearer service
I231.2	Circuit-mode 64 kbit/s, 8 kHz structured bearer service usable for speech

I464	Multiplexing, rate adaption and support of existing interfaces for restricted 64 kbit/s transfer capability
I465/V.120	Support by an ISDN of data terminal equipment with V-series type interfaces with provision for statistical multiplexing
I470	Relationship of terminal functions to ISDN
I500	General structure of the ISDN interworking Recommendations
I501	Service interworking
I510	Definitions and general principles for ISDN interworking
I511	ISDN-to-ISDN layer 1 internetwork interface
I515	Parameter exchange for ISDN interworking
I520	General arrangements for network interworking between ISDNs
I525	Interworking between networks operating at bit rates less than 64 kbit/s with 64 kbit/s-based ISDN and B-ISDN
I530	Network interworking between an ISDN and a public switched telephone network (PSTN)
I540/X321	General arrangements for interworking between Circuit-Switched Public Data Networks (CSPDNs) and ISDNs for the provision of data transmission services
I550/X325	General arrangements for interworking between Packet-Switched Public Data Networks (PSPDNs) and ISDNs for the provision of data transmission services
I555	Frame Relaying Bearer Service interworking
I560	Technical requirements to be met in providing the international telex service within an integrated services digital network
I570	Public/private ISDN interworking
I571	Connection of VSAT based private networks to the public ISDN
I580	General arrangements for interworking between B-ISDN and 64 kbit/s based ISDN
I581	General arrangements for B-ISDN interworking
I601	General maintenance principles of ISDN subscriber access and subscriber installation

I610	B-ISDN operation and maintenance principles and functions
I620	Frame relay operation and maintenance principles and functions
I630	ATM protection switching
I731	Types and general characteristics of ATM equipment
I732	Functional characteristics of ATM equipment
I751	Asynchronous transfer mode management of the network element view
V464	Rate adaption and support of existing interfaces for restricted 64 kbit/s transfer capability (to meet existing PCM transmissions in some countries outside the UK)

25.11.5 Videoconferencing

There are three groups of videoconferencing standards, covering the transmission media of ISDN, LAN and PSTN.

25.11.5.1 Videoconferencing over the ISDN (H320)

H100	Visual telephone systems
H110	Hypothetical reference connections for videoconferencing using primary digital group transmission
H120	Codecs for videoconferencing using primary digital group transmission
H130	Frame structures for use in the international interconnection of digital codecs for videoconferencing or visual telephony
H140	A multipoint international videoconference system
H200	Framework for Recommendations for audiovisual services
H221	Multiplexer video/audio user data
H222.0	Generic coding of moving pictures and associated audio information: Systems
H222.1	Multimedia multiplex and synchronization for audiovisual communication in ATM environments
H224	A real time control protocol for simplex applications using the H221 LSD/HSD/MLP channels
H226	Channel aggregation protocol for multilink operation on circuit-switched networks

H230	Control and indication
H231	Multipoint control units for audiovisual systems using digital channels up to 1920 kbit/s
H233	Confidentiality system for audiovisual services
H234	Encryption key management and authentication system for audiovisual services
H235	Security and encryption for H-Series multimedia terminals
H242	Videoconference protocol
H243	Multipoint connections
H244	Synchronized aggregation of multiple 64 or 56 kbit/s channels
H246	Interworking of H-Series multimedia terminals on GSTN and ISDN
H247	Multipoint extension for broadband audiovisual communication systems and terminals
H261	Video coding using BCH and forward error correction G711, G722 and G728 audio standards are also included within H320
H225	Packetizer protocol
H245	End-to-end control
H261/263	Video coding using BCH and forward error correction
H262	Generic coding of moving pictures and associated audio information: Video
H281	A far end camera control protocol for videoconferences using H224
H310	Broadband audiovisual communication systems and terminals
H320	Narrow-band visual telephone systems and terminal equipment
H321	Adaptation of H320 visual telephone terminals to B-ISDN environments
H322	Visual telephone systems and terminal equipment for local area networks
H323	Packet-based multimedia communications systems
H324	Terminal for low bit-rate multimedia communication
H331	Broadcasting type audiovisual multipoint systems and terminal equipment
H332	H323 extended for loosely coupled conferences
Q931	Call signalling

RAS	Gate keeper signalling
T120	User data, file transfer, etc.
H223	Multiplexer video/audio user data
H245	Videoconference protocol
H263	Video coding
T120	User data, file transfer, etc.

25.11.6 Telex

| R20 | Telex: Single-channel voice frequency |
| R111 | Telex: Multiplexing of 300 band signals |

25.11.7 Numbering and routing

E100	Definitions of terms used in international telephone operation
E104	International telephone directory assistance service and public access
E105	International telephone service
E110	Organization of the international telephone network
E111	Extension of international telephone services
E115	Computerized directory assistance
E123	Notation for national and international telephone numbers
E161	Arrangement of digits, letters and symbols on telephones and other devices for network access
E164	The international public telecommunication numbering plan
E164 Suppl. 1	Alternatives for carrier selection and network identification
E164 Suppl. 2	Number Portability
E166/X122	Numbering plan interworking for the E164 and X121 numbering plans
E167	ISDN network identification codes
E170	Traffic routing
E171	International telephone routing plan
E172	ISDN routing plan
E173	Routing plan for interconnection between public land mobile networks and fixed terminal networks
E174	Routing principles and guidance for Universal Personal Telecommunications (UPT)
E175	Models for international network planning
E177	B-ISDN routing

354

25.11.8 NETs

NET 1	Equipment connected to a public switched network with the CCITT X21 interface
NET 2	Equipment connected to the CCITT X25 interface to a packet switched network
NET 3	Equipment connected to an ISDN network, using the basic access 2B + D channels
NET 4	Public access to the PSTN for access to a single exchange line with simple signalling for terminal equipment such as telephones, modems, facsimile, teletext, etc. The specifications cover safety, call establishment, power levels and protocols
NET 20–25	Characteristics of modems
NET 30	Requirements for group 3 facsimile equipment
NET 31	Requirements for group 4 facsimile on ISDN
NET 33	Requirements for telephony terminal equipment connected to ISDN

25.11.9 Digital encoding of analogue signals

G701	Pulse code modulation (PCM) terms
G702	Digital hierarchy bit rates
G703	Bipolar line code for transmission of 64 kbit/s and higher data rates
G704	Framing of multiplexed data on multiplexed circuits (e.g. E1)
G706	Frame alignment and cyclic redundancy check (CRC) procedures (defined in G704)
G707	Network node interface for the synchronous digital hierarchy (SDH)
G711	3.1 kHz bandwidth audio to digital conversion, 56/64 kbit/s
G712	Transmission performance characteristics of PCM channels
G722	7 kHz bandwidth audio to digital conversion, 48/56/64 kbit/s
G723	3.1 kHz bandwidth audio to digital conversion, 5.3/6.4 kbit/s
G723.1	Dual rate speech coder for multimedia communications transmitting at 5.3 and 6.4 kbit/s
G724	48-channel low bit rate, encoding primary multiplex operating at 1544 kbit/s

26 Useful information

26.1 Abbreviations and acronyms

AAL-5	ATM Adaptation Layer, type 5 (layer-2 encapsulation; ATM)
AC	Alternating current
Ack	Acknowledgement
ADM	Add/drop multiplexes
ADPCM	Adaptive differential pulse code modulation
ADSL	Asymmetric digital subscribes line
ALU	Arithmetic logic unit
AM	Amplitude modulation
AMI	Alternate mark inversion
AMPS	Advanced mobile phone system
ANSI	American National Standards Institute
API	Application Programming Interface
APS	Automatic Protection Switching (SONET/SDH)
AQL	Acceptable quality level
ARP	Address Resolution Protocol
AS	Autonomous System
ASCII	American Standard Code for Information Interchange
ASIC	Application Specific Integrated Circuit
ASP	Application Service Provider
ATM	Asynchronous transfer mode
AUC	Authentication Centre (GSM)
BABT	British Approvals Board for Telecommunications
BCD	Binary-coded decimal
BER	Bit Error Rate
BGP	Border Gateway Protocol
BH	Busy hour
BOC	Bell Operating Company
BP	Band-pass (filter)
BPDU	Bridge Protocol Data Unit
BPS	Bits per second
BRI	Basic rate interface
BS	British Standard
BSI	British Standards Institution

BT	British Telecom
BVH	Buried via hole
BW	Bandwidth
CAI	Common air interface
CAMEL	Customised Applications for Mobile Enhanced Logic
CAS	Channel associated signalling
CBQ	Class-Based Queuing (QoS)
CBR	Constant Bit Rate (QoS)
CBT	Core-Based Tree (multicast routing)
CCIR	International Radio Consultative Committee
CCITT	International Telegraph & Telephone Consultative Committee
CCS	Common-channel signalling
CD	Collision detection
CDMA	Code division multiple access
CDPD	Cellular Digital Packet Data
CEC	Commission of European Communities
CEN	European Committee for Normalization
CENELEC	European Committee for Electrotechnical Normalization
CEPT	European Conference of Postal and Telecommunications Administrations
CHAP	Challenge Handshake Authentication Protocol
CIDR	Classless Internet Domain Routing
CISPR	Comite International Special Des Perturbations Radioelectriques (Radio Interference standards body)
CLI	Calling Line Identity
CMI	Coded mark inversion
CMOS	Complementary metal oxide semiconductor
CO	Central Office (telephone exchange)
Codec	Coder/decoder
CoP	Code of Practice
CoS	Class of Service
CPE	Customer premises equipment
CPU	Central processor unit
Cr	Chemical symbol for chromium
CRC	Cyclic redundancy check
CR-LDP	Constraint-based Routing Label Distribution Protocol (MPLS)

CRM	Customer Relations Management
CRT	Cathode ray tube
CSA	Canadian Standards Association
CSMA	Carrier Sense Multiple Access
CT	Cordless telephone
CTCSS	Continuous tone controlled signalling system
CTS	Conformance testing services
CW	Continuous wave
DACS	Digital Access Concentrator Switch
DASS	Digital access signalling system
dB	Decibel
DBE	Digital baseband equipment
DC	Direct current
DCE	Data circuit equipment
DDI	Direct dialling in
DDSN	Digital derived services network
DECT	Digital European cordless telephone
DHCP	Dynamic Host Configuration Protocol
DiffServ	Differentiated Services (IETF)
d.i.l.	Dual-in-line
DIV	Data in voice
DLC	Digital Loop carrier
DLCI	Data Link Channel Identifier (Frame Relay)
DLE	Digital local exchange
DLSW	Data Link Switching Protocol
DMSU	Digital main switching unit
DMT	Discrete Multi-Tone
DOV	Data over voice
DPCM	Differential pulse code modulation
DPNSS	Digital private network signalling system
DPSK	Differential phase shift keying
DR	Designated Router
DS0	Digital Signal level 0 (a 64 Kbps signal)
DS1	Digital Signal level 1 (a 1.544 Mbps signal, T1 carrier)
DS2	Digital Signal level 2 (a 6.312 Mbps signal, T2 carrier)
DS3	Digital Signal level 3 (a 44.76 Mbps signal, T3 carrier)
DS4	Digital Signal level 4 (a 274.176 Mbps signal)
DSB	Double sideband

DSC	District switching centre
DSCP	Differentiated Service Code Point
DSL	Digital Subscriber Line
DSLAC	Digital Subscriber Line Access Concentrator
DSLAM	Digital Subscriber Line Access Multiplexer
DSn	Digital Signal n
DTE	Data terminating equipment
DTI	Department of Trade & Industry
DTMF	Dual-tone multifrequency
DUP	Data user part
DUT	Device (or system) Under Test
DVMRP	Distance Vector Multicast Routing Protocol
DWDM	Dense Wavelength Division Multiplexing
DWMT	Discrete Wavelet Multi-Tone
EAP	Extensible Authentication Protocol
EBCDIC	Extended binary-coded decimal interchange code
EC	European Commission
ECITC	European Committee for Information Technology Certification
ECL	Emitter coupled logic
ECMA	European Computer Manufacturers Association
EDGE	Enhanced Data for GSM Evolution
EDI	Electronic data interface
EEC	European Economic Community
EF	Expedited forwarding (QoS)
EFTA	European Free Trade Association
EFWS	Electronic four-wire switch
EGP	Exterior Gateway Protocol (routing protocol)
EGRPS	Enhanced General Packet Radio Service
EIGRP	Enhanced Interior Gateway Routing Protocol (Cisco proprietary)
EIRP	Equivalent isotropic radiated power
em	Electromagnetic
EMC	Electromagnetic compatibility
EMI	Electromagnetic interference
EN	European Norm (regulatory)
EN	European Standard specification
ENV	Draft European Standard specification
EO	End office
EPD	Early Packet Discard (ATM)

ERO	Explicit Route Object
es	Electrostatic
ESA	European Space Agency
ESCD	Enhanced High-Speed Circuit Switched Data
ESPA	European Selective Paging Manufacturers Association
ESPRIT	European Strategic Program for Research and Development in Information Technologies
ETS	European Telecommunications Standard
ETSI	European Telecommunications Standards Institute
EUTELSAT	European Telecommunications Satellite Organization
F	Farad – unit of capacitance
FAW	Frame alignment word
FAX	Facsimile communication
FCC	Federal Communications Commission
FCS	Frame check sequence
FDD	Frequency Division Duplex
FDDI	Fibre distributed data interface
FDM	Frequency division multiplex
FDMA	Frequency-division multiple access
FEC	Forward error control
FEP	Front-end processor
FET	Field-effect transistor
FFSK	Fast frequency shift keying
FIFO	First in-first out
FIR	Finite impulse response
FISU	Fill-in signal unit
FM	Frequency modulation
FOCC	Forward control channel
FOTS	Fibre optic transmission systems
FPGA	Field-Programmable Gate Array
FR	Frame Relay
FSK	Frequency shift keying
FSVFT	Frequency shift multichannel voice-frequency telegraphy
FTP	File transfer protocol
FVC	Forward voice channel
FX	Foreign exchange

G.DMT	G-series ITU standards for ADSL using DMT
Gbps	Gigabits per second
GFR	Guaranteed frame Rate (ATM)
GGP	Gateway–Gateway Protocol LAN Local Area Network
GGSN	Gateway GPRS Support Node
GHz	$Hz \times 10^9$
GMSK	Gaussian minimum shift keying
GPRS	General Packet Radio Service
GSC	Group switching centre
GSM	Global System for Mobile
GSR	Gigabit Switch Router (Cisco proprietary)
GTO	Gate turn-off (transistor switch)
GUI	Graphical User Interface
H	Henry – unit of inductance
HDB	High-density Bipolar
HDLC	High-level data link control
HDSL	High-rate digital subscriber line
HDTV	High-Definition Television
HF	High frequency (3–30 MHz range)
HLR	Home Location Register (cellular)
HP	High-pass (filter)
HSCSD	High-Speed Circuit-Switched Data
HTML	Hyper-Text Mark-up Language
HTTP	Hyper-Text Transfer Protocol
Hz	(Hertz) Cycles/second
I-ETS	Interim European Telecommunications Standard
IA	International alphabet
IAD	Integrated Access Device
ICMP	Internet Control Message Protocol
ID	Inner diameter
IDC	Insulation displacement connector
IDN	Integrated digital network
IEC	Inter exchange carrier
IEC	International Technical Commission
IEEE	Institute of Electrical and Electronics Engineers
IETF	Internet Engineering Task Force
IF	Intermediate frequency
IFRB	International Frequency Registration Board
IGMP	Internet Group Management Protocol

IGP	Interior Gateway Protocol
IGRP	Interior Gateway Routing Protocol (Cisco proprietary)
IIH	IS-IS Hello (IS-IS)
IIR	Infinite impulse response
IN	Intelligent network
INMARSAT	Internation Maritime Satellite Organization
IntServ	Integrated Services (IETF)
IP	Internet protocol
IPCP	Ip Control Protocol (for PPP)
IPv4	Internet Protocol, Version 4
IPv6	Internet Protocol, Version 6
ISDN	Integrated services digital network
ISH	Intermediate System Hello (IS-IS)
IS-IS	Intermediate System-Intermediate System
ISO	International Standards Organization
ISP	Internet Service Provider
ISPBX	Integrated services private branch exchange
IT	Information technology
ITSTC	Information Technology Steering Committee
ITU	International Telecommunications Union
IVDW	Integrated voice data workstation
JCW	Justification control word
JDT	Justification time slot
Kbps	Kilobits per second
KHz	Hz $\times 10^3$
L2F	layer 2 Forwarding (for VoIP)
L2TP	Layer 2 Tunnelling Protocol (for VoIP)
LAN	Local-area network
LAP	Link access protocol
LAPB	Link access protocol balanced
LATA	Local access transport area
LCD	Liquid-crystal display
LCP	Link Control Protocol (for PPP)
LCS	Line control subsystem
LDAP	Lightweight Directory Access Protocol (X.500)
LDP	Label Distribution Protocol (MPLS)
LDR	Light-dependent resistor
LE	Local exchange
LEC	Local exchange carrier
LED	Light-emitting diode

LER	Label Edge Router (MPLS)
LLC	Logical Link Control
LMDS	Local Multi-point Distribution Service
LOF	Loss of Frame
LOGOS	Programming language for LCS and EFWS
LOP	Loss of Pointer
LOS	Loss of Signal
LP	Low-pass (filter)
LRC	Longitudinal redundancy check
LS	Link State (OSPF, IS-IS)
LSA	Link State Advertisement (OSPF)
LSI	Large-scale integration
LSP	Label Switch Path (MPLS)
LSP	Link State PDU (IS-IS)
LSR	Label Switch Router (MPLS)
LSSU	Link status signal unit
LTE	Line Terminal Equipment
LW	Long-wave
MAC	Media Access Control (Ethernet)
MAPOS	Multiple Access Protocol Over SONET/SDH
MBGP	Multicast Border Gateway Protocol
Mbps	Megabits per second
MDF	Main Distribution Frame
MDSN	Managed data service network
MGCP	Media Gateway Control Protocol (for VoIP)
MHz	Hz $\times$ 10^6
MIB	Management Information Base
MII	Media Independent Interface
MLB	Multilayer board
Modem	Modulator/demodulator
MOSFET	Metal oxide semiconductor FET
MOSPF	Multicast Open Shortest Path First
MPEG	Moving Picture Experts Group (video compression)
MPLS	Multi-Protocol Label Switching (Cisco)
MPOA	Multi-Protocol Over ATM
MPSK	Multiphase shift keying
MSA	Metropolitan statistical area
MSC	Main switching centre
MSC	Mobile Switching Centre
MSDP	Multicast Source Distribution Protocol

MSI	Medium-scale integration
MSN	Multiple subscriber number
MSRC	Metrology and Standards Requirements Committee
MSU	Message signal unit
MTBF	Mean time between failures
MTG	Multitone generator
MTP	Message transfer part
MTSO	Mobile telephone switching offices
MTTR	Mean time to repair
MTU	Maximum Transmit Unit (frame size)
MUX	Multiplexer
NAK	Negative acknowledgement
NATA	North American Telecommunications Association
NBMA	Non·Broadcast Multi-Access (OSPF)
NDC	Standard National Number
NEL	National Engineering Laboratory
NET	European Telecommunications Standard
NEXT	Near-end cross-talk
nF	$F \times 10^{-9}$
NF	Noise figure
nH	nanohenries($h \times 10^{-9})$)
Ni	Chemical symbol for nickel
NIC	Network Interface Card
NLPID	Network Layer Protocol Identification
NLRI	Network Layer Reach-ability Information
nm	Metres $\times 10^{-9}$
NNTP	Network News Transfer Protocol
NPL	National Physical Laboratory
NRTLIC	Nationally Recognized Testing Laboratory (Canada)
NRZ	Non-return to zero
NSC	Network Service Centre
NSP	Node – Switch Protocol
NT	Network termination
NTC	Negative temperature coefficient
NTP	Network Time Protocol
NWML	National Weights & Measures Laboratory
OC	Optical carrier
OC-n	Optical carrier n (where n is the level of SONET optical hierarchy: 1, 3, 9, etc.)

OCR	Optical character recognition
OD	Outer diameter
Oftel	Office of Telecommunications
ONA	Open network architecture
ONP	Open network provision
OSI	Open Systems Interconnection
OSPF	Open Shortest Path First
OSS	Operational Support System
PABX	Private automatic branch exchange
PAD	Packet assembler/disassembler
PAL	Phased alternate line
PAL	Programmable array logic
PAM	Pulse amplitude modulation
PAP	Password Authentication Protocol
PBX	Private branch exchange
p.c.b.	Printed circuit board
PCM	Pulse code modulation
PCN	Personal communications network
PCO	Public call office
PDH	Pleisiochronous Digital Hierarchy (transmission)
PDU	Protocol Data Unit (such as a packet)
pF	$F \times 10^{-12}$
PFQ	Per-Flow Queuing (QoS)
PHB	Per-Hop Behaviour (QoS)
PIG	PSTN Internet Gateway
PIM	Protocol Independent Multicast
PIN	P – intrinsic-n junction
PLD	Programmable logic device
PLL	Phase-locked loop
PM	Pulse modulation
PMR	Private mobile radio
pn	A junction of positive and negative minority carrier semiconductor materials
POCSAG	Post Office Code Standardization Advisory Group
POP	Point of presence
POS	Packet Over SONET/SDH
POTS	Plain Old Telephone Service
PPD	Partial Packet Discard (ATM)
PPM	Parts Per Million
ppm	Parts per million
PPP	Point to Point Protocol

PRI	Primary rate interface
PROM	Programmable ROM
ps	Seconds $\times 10^{-12}$
PSE	Packet switch exchange
PSK	Phase shift keying
PSN	Public switched network
PSNP	Partial Sequence Numbers PDU (IS-IS)
PSO	Public service operator
PSPDN	Packet switched public data network
PSS	Packet switched service
PSTN	Public switched telephone network
PTC	Positive temperature coefficient
PTH	Plated through hole
PTO	Public telephone (network) operator
QAM	Quadrature amplitude modulation
QoS	Quality of Service
QPSK	Quaternary phase shift keying
Q	Quality factor of inductance
RACE	Research & Development in Advanced Communications in Europe
RAM	Random access memory
RARP	Reverse Address Resolution Protocol
RAS	Remote Access Service
RBOC	Regional Bell Operating Company
RECC	Reverse control channel
RED	Random Early Detection (QoS)
REI	Remote Error Indication
RF	Radio frequency
RFC	Request for Comment (IETF)
RFI	Radio frequency interference
RIP	Routing Information Protocol
RMON	Remote Monitoring Protocol
RMS	Root mean square
ROM	Read-only memory
RP	Rendezvous Point
RPF	Reverse Path Forwarding
RPT	Rendezvous Point-based Tree
RSVP	Resource ReSerVation Protocol
RTCP	Real-time Transport Protocol Control Protocol (for VoIP)
RTP	Real-time Transport Protocol (for VoIP)

RTT	Round Trip Time (QoS)
RVC	Reverse voice channel
Rx	Receive
SAFI	Subsequent Address Family Identifier (BGP·4)
SAR	Segmentation and Re-assembly
SAW	Surface acoustic wave
Sb	Chemical symbol for antimony
SCOP	System control and operating position
SCP	System control processor
SCR	Silicon-controlled rectifier
SCVF	Single-channel voice frequency
SDH	Synchronous digital hierarchy
SDLC	Synchronous data link control
SDP	Session Description Protocol (for VoIP)
SDS	Space-division switching
SDSL	Synchronous Digital Subscriber Line
SEMKO	Swedish Standards Authority
SES	Ship earth station
SGCP	Simple Gateway Control Protocol (for VoIP)
s.i.l.	Single in line
SIM	Subscriber Identification Module (GSM)
SIP	Session Initiation Protocol (for VoIP)
SLA	Service Level Agreement
SMS	Short Message Service
SMTP	Simple Mail Transfer Protocol
Sn	Chemical symbol for tin
SN	Subscriber number
SNAP	Sub-Network Access Protocol
SNMP	Simple Network Management Protocol
S/N	Signal/noise ratio
SONET	Synchronous optical network
SP	Signalling point
SPC	Stored program control
SPE	Synchronous Payload Envelope (SONET)
SPF	Shortest Path First
SPT	Shortest Path Tree
SRBP	Synthetic resin-bonded paper
SS7	Signalling System No.7
SSAC	Signalling system alternating current
SSB	Single sideband
SSBSC	Single-sideband suppressed carrier

SSP	Switch–Switch Protocol
SSR	Source select register
SSS	Self-Synchronous Scrambler (POS)
STM	Synchronous Transport Module
STP	Signal Transfer Point (SS7)
STS	Space time space (digital switching)
STS	Synchronous Transport Signal (SDH)
T1	1.544 Mbps TDM system
T2	6.312 Mbps TDM system
T3	44.76 Mbps TDM system
TACS	Total access communications system
TASI	Time-assigned speech interpolation
TAT	Transatlantic telephone
TC	Temperature coefficient
TCL	Tool Command Language
TCLSH	Tool Command Language shell
TCP	Transmission Control Protocol
TDD	Time Division Duplex
TDM	Time division multiplex
TDMA	Time division multiple access
TDS	Time-division switching
TE	Terminal equipment
TIA	Telecommunications Industry Association
TMA	Telecommunications Managers Association
TNA	Telex network adaptor
TOS	Type of Service (IP)
TRAC	Technical Recommendations Advisory Committee
TRS	Transmitter repeater station
TS	Time slot
TSS	Testing support services
TST	Time space time (digital switching)
TTL	Time To Live (IP)
TTL	Transistor–transistor logic
TUP	Telephone user part
TWT	Travelling wave tube
Tx	Transmit
μF	$F \times 10^{-6}$
μm	Metres $\times 10^{-6}$
UART	Universal asynchronous receiver/transmitter
UDP	User Datagram Protocol

UL	Underwriters Listed (USA Standard)
UMTS	Universal Mobile Telecommunications System
USB	Universal Serial Bus
VAN	Value-added network
V_{be}	Voltage base-emitter
VBR	Variable Bit Rate (ATM)
VCC	Virtual Circuit Connection (ATM)
VCI	Virtual Channel Identifier (ATM)
VCO	Voltage-controlled oscillator
VDE	German Standards Authority
VDR	Voltage-dependent resistor
VDSL	Very-high-rate digital subscriber line
VDU	Visual display unit
VHF	Very high frequency
VOD	Video On Demand
VoDSL	Voice over Digital Subscriber Line
VoIP	Voice over IP
VPI	Virtual Path Identifier (ATM)
VPN	Virtual Private Network (ATM)
VRC	Vertical redundancy check
VSB	Vestigial sideband
VSWR	Voltage standing wave ratio
WAN	Wide-area network
WAP	Wireless Application Protocol
W-CDMA	Wide-band Code-Division Multiple Access
WDM	Wavelength Division Multiplexing (optical)
WFQ	Weighted Fair Queuing
WML	Wireless mark-up language
WRED	Weighted Random Early Detection
WWW	World Wide Web
XDSL	family of DSL standards – data over twisted-pair cables

26.2 Formulae relevant to telecommunications

26.2.1 Exponential, circular and hyperbolic functions

The base of natural or Napierian logarithms is 'e':

$$A = e^x \quad \log_e A = x$$

$$e^x = 1 + x + \frac{x^2}{2!} + \frac{x^3}{3!} + \frac{x^4}{4!} + \cdots$$

$$e^{-x} = 1 - x + \frac{x^2}{2!} - \frac{x^3}{3!} + \frac{x^4}{4!} - \cdots$$

$$e = 1 + 1 + \frac{1}{2!} + \frac{1}{3!} + \frac{1}{4!} + \cdots = 2.178\,28$$

Differentiating the series for e^x produces an identical series:

$$\frac{d}{dx}e^x = e^x \qquad \int e^x = e^x$$

Circular functions are based on a circle of radius a:

$$x^2 + y^2 = a^2$$

where x and y are the coordinates on the X and Y axes (Cartesian coordinates) or

$$a < \theta$$

where θ is the angle between the radius vector and the X axis (polar coordinates);

$$\tan\theta = \frac{y}{x}$$

θ may be expressed in radians (where θ = subtended arc/radius a) or degrees:

$$1 \text{ radian} = 360°/2\pi = 57.3°$$

$$\sin\theta = \frac{x}{a} = \theta - \frac{\theta^3}{3!} + \frac{\theta^5}{5!} - \cdots$$

$$\cos\theta = \frac{y}{a} = 1 - \frac{\theta^2}{2!} + \frac{\theta^4}{4!} - \cdots$$

$$\tan\theta = \frac{\sin\theta}{\cos\theta}$$

If θ varies with time, $\theta = \omega t$, where ω is the angular frequency in radians per second and t is the time in seconds. The vector then rotates at an angular velocity of ω, and sine and cosine become cyclic time functions with a repetition frequency of $f = \omega/2\pi$ cycles per second.

Cos θ leads sin θ by $\pi/2$ radians, representing a phase shift of $+\pi/2$:

$$\cos\theta = \sin\left(\theta + \frac{\pi}{2}\right)$$

372

Operator $(j)^2 = -1$ advances vector by π radians (reversing polarity):

$$j = \sqrt{-1}$$

and advances vector by $\pi/2$ radians.

$$\cos\theta = j\,\sin\theta \quad \sin\theta = -j\,\cos\theta$$

Hyperbolic functions (*Figure 26.1*) are based on the rectangular hyperbola having the equation $x^2 - y^2 = a^2$:

$$\sinh\theta = \frac{y}{a} = \theta + \frac{\theta^3}{3!} + \frac{\theta^5}{5!} + \cdots = \frac{e^\theta - e^{-\theta}}{2}$$

$$\cosh\theta = \frac{x}{a} = 1 + \frac{\theta^2}{2!} + \frac{\theta^4}{4!} + \cdots = \frac{e^\theta + e^{-\theta}}{2}$$

$$\tanh\theta = \frac{\sinh\theta}{\cosh\theta} = \frac{e^\theta - e^{-\theta}}{e^\theta + e^{-\theta}}$$

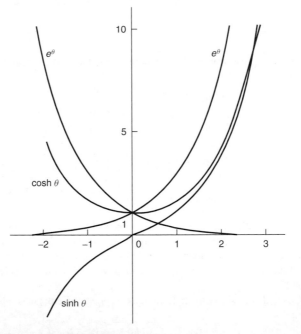

Figure 26.1 Exponential functions

Other hyperbolic and circular identities are:

$$\cosh j\theta = \cos\theta$$
$$\sinh j\theta = j\,\sin\theta$$
$$\tanh j\theta = j\,\tan\theta$$
$$\cosh\theta = \cos j\theta$$
$$j\,\sinh\theta = \sin j\theta$$
$$j\,\tanh\theta = \tan j\theta$$
$$\operatorname{cosec}\theta = 1/\sin\theta$$
$$\sec\theta = 1/\cos\theta$$
$$\cot\theta = 1/\tan\theta$$
$$\operatorname{cosech}\theta = 1/\sinh\theta$$
$$\operatorname{sech}\theta = 1/\cosh\theta$$
$$\coth\theta = 1/\tanh\theta$$

Sum and difference formulae:

$$\cos^2\theta + \sin^2\theta = 1$$
$$\sec^2\theta - \tan^2\theta = 1$$
$$\cosh^2\theta - \sinh^2\theta = 1$$
$$\operatorname{sech}^2\theta + \tanh^2\theta = 1$$
$$\tan(A \pm B) = \frac{\tan A \pm \tan B}{1 \pm \tan A \tan B}$$
$$\tanh(A \pm B) = \frac{\tanh A \pm \tanh B}{1 \pm \tanh A \tanh B}$$
$$\sin(A \pm B) = \sin A \cos B \pm \cos A \sin B$$
$$\cos(A \pm B) = \cos A \cos B \pm \sin A \sin B$$
$$\sinh(A \pm B) = \sinh A \cosh B \pm \cosh A \sinh B$$
$$\cosh(A \pm B) = \cosh A \cosh B \pm \sinh A \sinh B$$
$$\sinh(A \pm jB) = \sinh A \cos B \pm j \cosh A \sin B$$
$$\cosh(A \pm jB) = \cosh A \cos B \pm j \sinh A \sin B$$

$$\sin A \pm \sin B = 2 \sin \left(\frac{A \pm B}{2} \right) \cos \left(\frac{A \pm B}{2} \right)$$

$$\cos A + \cos B = 2 \cos \left(\frac{A + B}{2} \right) \cos \left(\frac{A - B}{2} \right)$$

$$\cos A - \cos B = -2 \sin \left(\frac{A + B}{2} \right) \sin \left(\frac{A - B}{2} \right)$$

$$\sin A \sin B = -\tfrac{1}{2}[\cos(A + B) - \cos(A - B)]$$

$$\cos A \cos B = \tfrac{1}{2}[\cos(A + B) + \cos(A - B)]$$

$$\sin A \cos B = \tfrac{1}{2}[\sin(A + B) + \sin(A - B)]$$

$$\cos A \sin B = \tfrac{1}{2}[\sin(A + B) - \sin(A - B)]$$

Differentials and integrals:

$$\frac{d}{dx} \sin ax = a \cos ax$$

$$\frac{d}{dx} \sinh ax = a \cosh ax$$

$$\frac{d}{dx} \cos ax = -a \sin ax$$

$$\frac{d}{dx} \cosh ax = a \sinh ax$$

$$\frac{d}{dx} \tan ax = a \sec^2 ax$$

$$\frac{d}{dx} \tanh ax = a \operatorname{sech}^2 ax$$

$$\frac{d}{dx} e^{ax} = a\, e^{ax}$$

$$\frac{d}{dx} \log_e (ax) = \frac{a}{x}$$

$$\frac{d}{dx} a^x = a^x \log_e a$$

$$\frac{d(uv)}{dx} = u\frac{dv}{dx} + v\frac{du}{dx}$$

$$\frac{d}{dx} \left(\frac{u}{v} \right) = \frac{v(du/dx) - u(dv/dx)}{v^2}$$

$$\int \sin ax \, dx = -\frac{1}{a} \cos ax$$

$$\int \sinh ax \, dx = \frac{1}{a} \cosh ax$$

$$\int \cos ax \, dx = \frac{1}{a} \sin ax$$

$$\int \cosh ax \, dx = \frac{1}{a} \sinh ax$$

$$\int \tan ax \, dx = -\frac{1}{a} \log_e \cos ax$$

$$\int \tanh ax \, dx = \frac{1}{a} \log_e \cosh ax$$

$$\int \sin^2 x \, dx = \frac{1}{2} \left(x - \frac{1}{2} \sin 2x \right)$$

$$\int \cos^2 x \, dx = \frac{1}{2} \left(x + \frac{1}{2} \sin 2x \right)$$

$$\int \sin Ax \sin Bx \, dx = -\frac{\sin(A+B)x}{2(A+B)} + \frac{\sin(A-B)x}{2(A-B)}$$

$$\int \cos Ax \cos Bx \, dx = \frac{\sin(A+B)x}{2(A+B)} + \frac{\sin(A-B)x}{2(A-B)}$$

Rotating vectors: vector length a rotating anticlockwise at ω radians/s is represented by

$$a \, e^{j\omega t} \text{ or } a(\cos \omega t + j \sin \omega t)$$

Vector length a rotating clockwise at ω radians/s is represented by

$$a \, e^{-j\omega t} \text{ or } a(\cos \omega t - j \sin \omega t)$$

Rotating fields: fields $H \cos \omega t$ and $jH \sin \omega t$ (two fields at right angles and $\pi/2$ phase displaced) combine to produce a rotating field $H e^{j\omega t}$

26.2.2 Resistance (R), reactance (X), impedance (Z) of two-terminal networks

Reactance of capacitor (C farads):

$$X_C = \frac{1}{j\omega C} \text{ ohms}$$

Reactance of inductor (L henries):

$$X_L = j\omega L \text{ ohms}$$

Impedance:

$$Z = R + jX = \sqrt{R^2 + X^2} < \tan^{-1}(X/R) \text{ ohms}$$

Admittance:

$$Y = 1/Z \text{ siemens}$$

Impedances in series:

$$Z_T = Z_1 + Z_2 + Z_3 + \cdots$$

Impedances in parallel:

$$\frac{1}{Z_T} = \frac{1}{Z_1} + \frac{1}{Z_2} + \frac{1}{Z_3} + \cdots$$

$$\text{Current}(I \text{ amps}) = \frac{\text{EMF (volts)}}{\text{Impedance (ohms)}}$$

Energy stored in capacitor:

$$\tfrac{1}{2}CV^2 \text{ joules}$$

where V is voltage across capacitor.

Energy stored in inductor:

$$\tfrac{1}{2}LI^2 \text{ joules}$$

where I is current in amps.

Tuned circuit: resonance occurs when $\omega L = \dfrac{1}{\omega C}$, that is at a frequency

$$\omega_R = \frac{1}{\sqrt{LC}} \text{ or } f_R = \frac{1}{2\pi\sqrt{LC}}$$

Impedance of series tuned circuit at resonance:

$$R = \frac{\omega L}{Q}$$

Impedance of parallel tuned circuit at resonance:

$$\frac{(\omega L)^2}{R} = Q\omega L$$

where Q, the goodness factor of tuned circuit, is

$$\frac{\omega L}{R}$$

Bandwidth of tuned circuit between 3 dB attenuation frequencies is

$$\frac{\omega_R}{Q}$$

Total stored energy in tuned circuit at resonance:

$$\tfrac{1}{2}LI^2$$

where I is the peak current circulating.

Rate of energy loss:

$$\frac{1}{Q} \times \text{stored energy per radian}$$

Foster's Theorem: A two-terminal network composed of pure reactances only has reactance

$$X = AB\frac{(\omega^2 - \omega_1^2)(\omega^2 - \omega_3^2)(\omega^2 - \omega_5^2)\cdots}{(\omega^2 - \omega_2^2)(\omega^2 - \omega_4^2)(\omega^2 - \omega_6^2)\cdots}$$

where ω_1, ω_3, ω_5 are series resonant frequencies (zeros), ω_2, ω_4, ω_6 are parallel resonant frequencies (poles), A and B are constants dependent on the effective values of X at zero and infinite frequencies.

Skin effect: At high frequencies current flow is concentrated in outer skin of conductor, reducing effective resistance

$$\text{Effective skin depth} = \sqrt{\frac{\rho}{\pi\mu_a f}} \text{ metres}$$

where

ρ = resistivity of conductor (metre ohms)
   = 1.74×10^{-8} for copper at 20°C
μ_a = permeability of conductor ($\mu_a = \mu_0 = 1.257 \times 10^{-6}$
   for copper in SI units)
f = frequency (Hz)

For copper this gives

$$\text{Skin depth} = \frac{0.0662}{\sqrt{f}} \text{ metres}$$

26.2.3 Power, nepers and decibels

$$\text{Power dissipated } (P) = I^2R = \frac{V^2}{R} \text{ watts (joules/s)}$$

$$= I \times V \cos\theta$$

where θ is the phase angle between V and I.

Power ratio is expressed in decibels, a logarithmic unit:

Ratio of powers P_1 and $P_2 = 10 \log_{10}(P_2/P_1)$ dB

If P_1 and P_2 refer to similar impedances $(V_1/V_2 = I_1/I_2)$ ratio of voltage or current can be expressed as

$$\text{Voltage ratio} = 20 \log_{10}(V_2/V_1)\,\text{dB}$$

$$\text{Current ratio} = 20 \log_{10}(I_2/I_1)\,\text{dB}$$

If P_1, V_1, I_1 are the inputs to a network and P_2, V_2, I_2, the outputs, +dB represents gain and −dB loss. Losses and gains of tandem-connected networks of similar characteristic impedance can therefore be summed algebraically.

Voltage or current ratio can be expressed in nepers, a logarithmic unit based on Napierian logarithms:

$$\text{Gain(loss) in nepers} = \log_e(V_2/V_1) \text{ or } \log_e(I_2/I_1)$$

$$+1 \text{ neper} = \text{gain of } 2.718\,28 \text{ times}$$

$$1 \text{ neper} \equiv 8.686\,\text{dB}$$

26.2.4 Four-terminal networks (Figure 26.2)

$$Z_{IN} = Z_0 \text{ when } Z_L = Z_0$$

$$Z_{OUT} = Z_0 \text{ when } Z_S = Z_0$$

$$Z_{IN} = Z_{SC} \text{ when } Z_L = 0$$

$$Z_{IN} = Z_{0C} \text{ when } Z_L = \infty$$

$$Z_0 = \sqrt{Z_{SC}Z_{0C}} \quad \tan h\gamma = \sqrt{Z_{SC}/Z_{0C}}$$

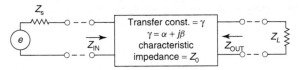

Figure 26.2 General four-terminal network

Reflection coefficient (receive):

$$r_R = \frac{Z_L - Z_0}{Z_L + Z_0} \quad (r_R = 0 \text{ when } Z_L = Z_0)$$

Reflection coefficient (send):

$$r_S = \frac{Z_S - Z_0}{Z_S + Z_0} \quad (r_S = 0 \text{ when } Z_S = Z_0)$$

Input impedance of four-terminal network:

$$Z_{IN} = Z_0 \frac{1 + e^{-2\gamma} r_R}{1 - e^{-2\gamma} r_R}$$

When $Z_L = 0$, $r_R = -1$, $Z_{IN} = Z_0 \tanh \gamma$

When $Z_L = \infty$, $r_R = +1$, $Z_{IN} = Z_0 \coth \gamma$

Return loss $= 20 \log_{10} r_R$

Insertion loss of four-terminal network:

$$e^{-\gamma} \frac{1 - r_R r_S}{1 - r_R r_S e^{-2\gamma}}$$

(equals $e^{-\gamma}$ if $Z_S = Z_R = Z_0$).
For symmetrical sections (*Figure 26.3*):

$$\text{T sections: } Z_0 = \sqrt{Z_1 Z_2 + \frac{Z_1^2}{4}}; \sinh \frac{\gamma}{2} = \sqrt{\frac{Z_1}{4Z_2}}$$

$$\pi \text{ sections: } Z_0 = \sqrt{\frac{4Z_1 Z_2^2}{Z_1 + 4Z_2}}; \sinh \frac{\gamma}{2} = \sqrt{\frac{Z_1}{4Z_2}}$$

$$\text{Lattice section: } Z_0 = \sqrt{Z_A Z_B}; \tanh \frac{\gamma}{2} = \sqrt{\frac{A_A}{Z_B}}$$

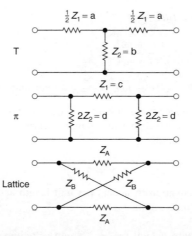

Figure 26.3 T, π and lattice

Table 26.1

Loss (dBs)	T (a)	T (b)	PI (c)	PI (d)
1	0.0575	8.668	0.1153	17.393
2	0.1147	4.305	0.2323	8.722
3	0.1708	2.838	0.3518	5.853
4	0.2263	2.097	0.477	4.418
5	0.28	1.645	0.6083	3.57
6	0.3323	1.339	0.7468	3.01
7	0.3823	1.117	0.8955	2.615
8	0.4305	0.9458	1.057	2.323
9	0.4762	0.8118	1.232	2.1
10	0.5195	0.7032	1.422	1.925
11	0.5605	0.612	1.634	1.785
12	0.5985	0.5362	1.865	1.672
13	0.6342	0.4712	2.122	1.577
14	0.6673	0.4155	2.407	1.499
15	0.698	0.3668	2.722	1.433
20	0.8182	0.202	4.95	1.222
25	0.8935	0.1127	8.873	1.119
30	0.8937	0.06332	15.81	1.065
35	0.9387	0.03558	28.107	1.036
40	0.9818	0.02	50.0	1.02
45	0.9888	0.01125	88.917	1.0113
50	0.9937	0.006325	158.1	1.0063

Table 26.1 gives normalized resistor values for attenuators using T and π networks. To de-normalize multiply all values by the source or load resistance.

26.2.5 Infinite line

$$Z_0 = \sqrt{\frac{R + j\omega L}{G + j\omega C}}$$

$$\gamma = \sqrt{(R + j\omega L)(G + j\omega C)} \text{ per unit length}$$

where R is resistance per unit length (ohms), G leakage per unit length (siemens), L inductance per unit length (henries) and, C capacitance per unit length (farads).

$$\gamma = \alpha + j\beta$$

where α is attenuation in nepers per unit length and β is phase shift in radians per unit length.

$$\text{Phase velocity} = \frac{\omega}{\beta} \quad \text{Group velocity} = \frac{d\omega}{d\beta}$$

If $\omega L \geq R$, $\omega C \geq G$

$$\beta = \omega\sqrt{LC} \text{ and } Z_0 = \sqrt{\frac{L}{C}}$$

$$\alpha = \frac{R}{2Z_0}$$

$$\text{Phase velocity} = \frac{1}{\sqrt{LC}}$$

26.2.6 Non-linear distortion

Transfer characteristic given by:

$$V_{OUT} = AV_{IN} + BV_{IN}^2 + CV_{IN}^3 + DV_{IN}^4 + \cdots$$

and is linear (no distortion) when $B, C, D, \ldots = 0$.

For applied input signal $V\cos\omega\, t$:

B term produces $\dfrac{BV^2}{2}(1 + \cos 2\omega t)\ldots$ second harmonic distortion

C term produces $CV^3\left(\frac{3}{4}\cos\omega t + \frac{1}{4}\cos 3\omega t\right)\ldots$ third harmonic distortion

For simultaneous applied signals of $V\cos\omega_1 t$ and $V\cos\omega_2 t$, B term produces

$$BV^2[1 + \cos(\omega_1 t - \omega_2 t) + \cos(\omega_1 t + \omega_2 t)$$
$$+ \tfrac{1}{2}(\cos 2\omega_1 t + \cos 2\omega_2 t)]$$

(sum and difference intermodulation products in addition to second harmonics) and C term produces

$$CV^3\left[\tfrac{5}{4}\cos\omega_1 t + \tfrac{5}{4}\cos\omega_2 t + \tfrac{1}{4}\cos 3\omega_1 t + \tfrac{1}{4}\cos 3\omega_2 t\right.$$
$$+ \tfrac{3}{4}\cos(2\omega_1 t - \omega_2 t) + \tfrac{3}{4}\cos(2\omega_1 t + \omega_2 t)$$
$$\left.+ \tfrac{3}{4}\cos(2\omega_2 t - \omega_1 t) + \tfrac{3}{4}\cos(2\omega_2 t + \omega_1 t)\right]$$

(fundamental, third harmonic and four sum and difference distortion components).

26.2.7 Thermal noise

Thermal noise $= V$ volts:

$$V^2 = 4kT\int_0^\infty R\,\mathrm{d}f$$

where k is Boltzmann's constant (1.37×10^{-23} watt seconds/degree), T absolute temperature, R resistive component of circuit impedance and f frequency in hertz.

For channel B Hz wide at 20°C with R constant over the band

$$V = \sqrt{1.64R.B \times 10^{-20}} \text{ volts RMS}$$

or

$$\text{noise power} = 1.64B \times 10^{-20} \text{ watts}$$

26.2.8 Amplitude modulation

Carrier $E_c \cos \omega_c t$ modulated to a modulation depth m by modulation signal $\cos \omega_m t$

$$v = E_c \cos \omega_c t(1 + m \cos \omega_m t)$$
$$= E_c[\cos \omega_c t + \tfrac{1}{2}m \cos(\omega_c + \omega_m)t + \tfrac{1}{2}m \cos(\omega_c - \omega_m)t]$$

Total bandwidth $= 2\omega_m$

$$\text{Percentage modulation} = 100m = \frac{E_{\text{MAX}} - E_{\text{MIN}}}{E_{\text{MAX}} + E_{\text{MIN}}} \times 100\%$$

where E_{MAX} and E_{MIN} are the maximum and minimum values of the modulated carrier.

Single-sideband suppressed carrier (SSBSC) has carrier and one sideband suppressed:

Bandwidth $= \omega_m$

26.2.9 Frequency and phase modulation

FM: For a carrier $E_c \sin \omega_c t$ and modulating signal $E_m \cos \omega_m t$

$$v = E_c \sin(\omega_c + \omega_D \cos \omega_m t)t$$

where ω_D = peak frequency deviation

PM: For a carrier $E_c \sin \omega_c t$ and modulating signal $E_m \sin \omega_m t$

$$v = E_c \sin(\omega_c t + \varphi_D \sin \omega_m t)$$

where φ_D = peak phase deviation (radians)

These formulae can be expressed by the single equation

$$v = E_c \sin(\omega_c t + m \sin \omega_m t)$$

where m = modulation index
$= \omega_D/\omega_m$ for FM φ_D for PM

Frequency spectrum required for FM or PM signal is given by

$$v = E_c\{J_0(m)\sin\omega_c t + J_1[\sin(\omega_c + \omega_m)t - \sin(\omega_c - \omega_m)t]$$
$$+ J_2(m)[\sin(\omega_c + 2\omega_m)t + \sin(\omega_c - 2\omega_m)t]$$
$$+ J_3(m)[\sin(\omega_c + 3\omega_m)t + \sin(\omega_c - 3\omega_m)t]$$
$$+ \cdots\}$$

where $J_0, J_1, J_2 \ldots$ are Bessel functions of the first kind with argument m and are shown in *Figure 26.4* for J_0 to J_4 (showing how the values of sidebands vary with m).

Sidebands $\omega_c \pm (m + 2)\omega_m$ and above are of negligible amplitude and effective bandwidth required is approximately

$$2\omega_m(m + 1)$$

The Bessel functions are cyclic, providing zeros at various values of m. The carrier passes through zero at the following m values:

$$2.41, 5.5, 8.65, 11.79, 14.93, 18.07, \ldots,$$

$$[18.07 + \pi(n - 6)]$$

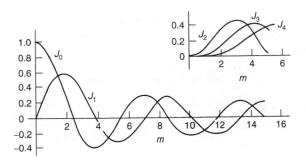

Figure 26.4   Bessel functions

26.2.10 Fourier series

For a repetitive waveform, repeating every t seconds,

$$f = 1/t \text{ Hz}$$
$$\omega = 2\pi/t \text{ radians/s}$$

Fourier analysis enables the waveform to be expressed in terms of a DC component and components at ω, 2ω, 3ω, ...

$$v = B_0 + A_1 \sin \omega t + B_1 \cos \omega t + A_2 \sin 2\omega t + B_2 \cos 2\omega t$$
$$+ A_3 \sin 3\omega t + B_3 \cos 3\omega t + A_4 \sin 4\omega t + B_4 \cos 4\omega t$$
$$+ \cdots$$

B_0 represents DC component

$\sqrt{(A_1^2 + B_1^2)} \sin(\omega t + \varphi_1)$ represents fundamental with

$\varphi_1 = \tan^{-1}(B_1/A_1)$

$\sqrt{(A_n^2 + B_n^2)} \sin(n\omega t + \varphi_n)$ represents component $n\omega$ with

$\varphi_n = \tan^{-1}(B_n/A_n)$

The coefficients B_0, A_n, B_n are given by

$$B_0 = \frac{\omega}{2\pi} \int_0^{2\pi/\omega} f(t)\, dt$$

$$A_n = \frac{\omega}{2\pi} \int_0^{2\pi/\omega} f(t) \sin n\omega t\, dt$$

$$B_n = \frac{\omega}{2\pi} \int_0^{2\pi/\omega} f(t) \cos n\omega t\, dt$$

For the typical waveforms shown in *Figure 26.5*:

Square wave:

$$v = \frac{2E}{\pi} \left[\sin \omega t + \frac{1}{3} \sin 3\omega t + \frac{1}{5} \sin 5\omega t + \cdots \right]$$

Triangular wave:

$$v = \frac{4E}{\pi^2} \left[\sin \omega t - \frac{1}{3^2} \sin 3\omega t + \frac{1}{5^2} \sin 5\omega t + \cdots \right]$$

Half cosine wave:

$$v = E \left[\frac{1}{\pi} + \frac{1}{2} \cos \omega t + \frac{2}{\pi} \left(\frac{\cos 2\omega t}{3} - \frac{\cos 4\omega t}{15} + \frac{\cos 6\omega t}{35} \right. \right.$$
$$\left. \left. - \frac{\cos 8\omega t}{63} + \cdots \right) \right]$$

Note that half cosine signal approximates to the rounded baseband signal sometimes used for digital transmission, and comparison with

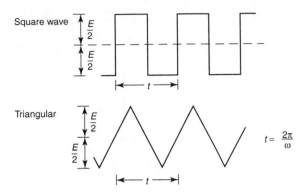

Figure 26.5 Waveforms

the square wave shows the much sharper drop-off in higher harmonic components.

26.2.11 Transmission in free space

Inductance of free space = 1.257×10^{-6} henries per metre

Capacitance of free space = 8.842×10^{-12} farads per metre

Impedance of free space = $\sqrt{L/C}$ = 377 ohms

Velocity of transmission in free space = $1/\sqrt{LC}$

$$= 3 \times 10^8 \text{ metres/s}$$

	mks (unrationalized)	S (rationalized)
Permeability of free space (μ_0)	10^{-7}	1.257×10^{-6}
Dielectric constant of free space (K_0)	$\dfrac{1}{9\pi 10^9}$	8.842×10^{-12}

Note that μ_0 and k_0 are the same as the inductance and capacitance of free space in the SI system of units, but differ in all other systems (i.e. emcgs, escgs, and unrationalized ... mks).

26.2.12 Negative feedback

$$\text{Gain of feedback amplifier} = \frac{\mu}{1 + \mu B}$$

386

Table 26.2

Amplifier mode	FB from output	FB to input	Output impedance	Input impedance	Gain
Voltage to voltage	Shunt	Series	$Z_{OUT}/(1 + \mu\beta)$	$Z_{IN}(1 + \mu\beta)$	V_{OUT}/V_{IN}
Current to current	Series	Shunt	$Z_{OUT}(1 + \mu\beta)$	$Z_{IN}/(1 + \mu\beta)$	i_{OUT}/i_{IN}
Voltage to current	Series	Series	$Z_{OUT}(1 + \mu\beta)$	$Z_{IN}(1 + \mu\beta)$	i_{OUT}/V_{IN} amps/volt
Current to voltage	Shunt	Shunt	$Z_{OUT}/(1 + \mu\beta)$	$Z_{IN}/(1 + \mu\beta)$	V_{OUT}/i_{IN} volts/amp

where μ = forward gain of amplifier
β = loss in feedback network

Characteristics of amplifiers with negative feedback (FB) are given in Table 26.2.

26.3 International alphabet No. 2 and ASCII code

Binary numbers are in ascending sequence 0–127

26.3.1 IA2

A -	11000
B ?	10011
C :	01110
D who are you	10010
E 3	10000
F %	10110
G @	01011
H £	00101
I 8	01100
J Bell	11010
K (	11110
L)	01001
M .	00111
N ,	00110
O 9	00011
P 0	01101
Q 1	11101
R 4	01010
S '	10100
T 5	00001
U 7	11100
V =	01111
W 2	11001
X /	10111
Y 6	10101
Z +	10001
Carriage return	00010
Figures	11011
Letters	11111
Line feed	01000
Space	00100

Note: '1' corresponds to a punched hole in paper tape.

26.3.2 ASCII

Binary	ASCII
00000000	NUL
00000001	SOH
00000010	STX
00000011	ETX
00000100	EOT
00000101	ENQ
00000110	ACK
00000111	BEL
00001000	BS
00001001	HT
00001010	LF
00001011	VT
00001100	FF
00001101	CR
00001110	SO
00001111	SI
00010000	DLE
00010001	DC1
00010010	DC2
00010011	DC3
00010100	DC4
00010101	NAK
00010110	SYN
00010111	ETB
00011000	CAN
00011001	EM
00011010	SUB
00011011	ESC
00011100	FS
00011101	GS
00011110	RS
00011111	US
00100000	SP
00100001	!
00100010	"
00100011	#
00100100	$

00100101	%
00100110	&
00100111	'
00101000	(
00101001	)
00101010	*
00101011	+
00101100	,
00101101	−
00101110	.
00101111	/
00110000	0
00110001	1
00110010	2
00110011	3
00110100	4
00110101	5
00110110	6
00110111	7
00111000	8
00111001	9
00111010	:
00111011	;
00111100	<
00111101	=
00111110	>
00111111	?
01000000	@
01000001	A
01000010	B
01000011	C
01000100	D
01000101	E
01000110	F
01000111	G
01001000	H
01001001	I
01001010	J
01001011	K
01001100	L
01001101	M
01001110	N

01001111	O
01010000	P
01010001	Q
01010010	R
01010011	S
01010100	T
01010101	U
01010110	V
01010111	W
01011000	X
01011001	Y
01011010	Z
01011011	[
01011100	\
01011101	]
01011110	"
01011111	—
01100000	`
01100001	a
01100010	b
01100011	c
01100100	d
01100101	e
01100110	f
01100111	g
01101000	h
01101001	i
01101010	j
01101011	k
01101100	l
01101101	m
01101110	n
01101111	o
01110000	p
01110001	q
01110010	r
01110011	s
01110100	t
01110101	u
01110110	v
01110111	w
01111000	x

01111001	y
01111010	z
01111011	{
01111100	:
01111101	}
01111110	-
01111111	DEL

26.4 Wire colour code

Plastic insulation has the advantage of being able to accept colour dyes. Manufacturers now produce cables with colour coded insulated wires. Colours are expressed as a primary (base colour) with a secondary colour band. Thus 'white blue' is mostly white with blue bands. In contrast, 'blue white' is mostly blue with white bands. *Table 26.3* gives the colours for the first 30 pairs.

26.5 Telephone jack wiring in the UK

The edited description provided here by EPL Ltd. refers to the diagram in *Figure 26.6* (also provided by EPL Ltd.).

The socket and apparatus wiring is coded as follows:

BK	Black
W	White
G	Green
B	Blue
R	Red
WG	White with green banding
WB	White with blue banding
WO	White with orange banding
OW	Orange with white banding
NW	Blue with white banding
GW	Green with white banding

Note 1: The wiring to the socket is of a single core type which is secured into the socket terminals using an insulation displacement tool.
Note 2: The cable most often used to connect apparatus to the wall socket is a multi-cored tinsel type wire manufactured for its flexibility. With tinsel cable it is important that the correct tools are used when

Table 26.3 Wire colours in telecommunications cables

Pair	A leg	B leg
1	White Blue	Blue White
2	White Orange	Orange White
3	White Green	Green White
4	White Brown	Brown White
5	White Grey	Grey White
6	Red Blue	Blue Red
7	Red Orange	Orange Red
8	Red Green	Green Red
9	Red Brown	Brown Red
10	Red Grey	Grey Red
11	Black Blue	Blue Black
12	Black Orange	Orange Black
13	Black Green	Green Black
14	Black Brown	Brown Black
15	Black Grey	Grey Black
16	Yellow Blue	Blue Yellow
17	Yellow Orange	Orange Yellow
18	Yellow Green	Green Yellow
19	Yellow Brown	Brown Yellow
20	Yellow Grey	Grey Yellow
21	Pink Blue	Blue Pink
22	Pink Orange	Orange Pink
23	Pink Green	Green Pink
24	Pink Brown	Brown Pink
25	Pink Grey	Grey Pink
26	Purple Blue	Blue Purple
27	Purple Orange	Orange Purple
28	Purple Green	Green Purple
29	Purple Brown	Brown Purple
30	Purple Grey	Grey Purple

making terminations as the individual strands of wire are very difficult to solder satisfactorily.

From *Figure 26.6* it is clear that the UK wiring arrangement is quite different to that of most countries. Some countries, such as Ireland and New Zealand, have a similar wiring arrangement but use a different type of socket.

The principal differences between the UK and other countries is the incorporation of a voltage arrestor device and components (470 kΩ resistor and a 1.8 μF capacitor) into the wall socket. These components allow testing of the telephone line from the telephone exchange.

The voltage arrestor device is a gas discharge tube (GDT) component intended to short circuit the A-wire to the B-wire in the event

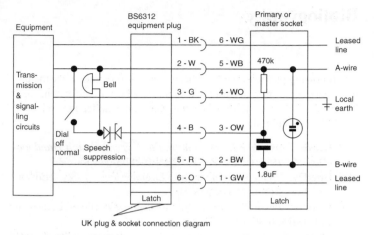

Figure 26.6   UK telephone socket wiring

of voltages exceeding approximately 250 V becoming present on the telephone line. This type of device is relatively slow acting and has been superseded by the installation of a polyswitch type of device in the line interface of most newly designed products.

The wall socket also contains a connection to the telecoms apparatus intended to suppress inductive spikes which are generated when loop-disconnect dialling into electro-mechanical exchanges which terminate the phone circuit with a relay coil.

Note: These exchanges are presently being phased out of operation, which, coupled with DTMF (MF4) detectors being a design feature of the replacement exchanges is resulting in the diminishing use of loop-disconnect dialling.

Bibliography

The following publications have been used for reference over the past few years. They may be useful to readers who wish to extend their knowledge in one of the specialisms that come under the 'telecommunications' umbrella.

[1] Reeve, *Subscriber Loop Signaling and Transmission Handbook (Digital)*, 1995 (IEEE Press) ISBN 0-7803-0440-3

[2] Brierley, *Telecommunications Engineering*, 1986 (Edward Arnold) ISBN 0-7131-3558-1

[3] Griffiths, *Local Telecommunications 2*, 1986 (Peter Peregrinus Ltd) ISBN 0-86341-080-4

[4] Mazda, *Telecommunications Engineer's Reference Book*, 1993 (Butterworth Heinemann) ISBN 0-7506-1162-6

[5] Dunlop & Smith, *Telecommunications Engineering*, 1989 (Van Nostrand Reinhold) ISBN 0-278-00082-7

[6] Carlson, *Communication Systems*, 1986 (McGraw-Hill) ISBN 0-07-100560-9

[7] Schweber, *Electronic Communication Systems*, 1996 (Prentice Hall) ISBN 0-13-373226-6

[8] McDonald, *Fundamentals of Digital Switching*, 1990 (Plenum) ISBN 0-306-43347-8

[9] Atkins & Norris, *Total Area Networking*, 1995 (Wiley) ISBN 0-471-95480-2

[10] Griffiths, *ISDN Explained*, 1992 (Wiley) ISBN 0-471-93480-1

[11] Hills, *Telecommunications Switching Principles*, 1979 (MIT Press) ISBN 0-262-08092-3

[12] Bigelow, *Understanding Telephone Electronics*, 1997 (Newnes) ISBN 0-7506-9944-2

[13] Feher, *Wireless Digital Communications*, 1995 (Prentice Hall) ISBN 0-13-098617-8

[14] Sklar, *Digital Communications*, 1988 (Prentice Hall) ISBN 0-13-212713-X

[15] Ronayne, *The Integrated Services Digital Network*, 1987 (Pitman) ISBN 0-273-02677-1

[16] Ronayne, *Introduction to Digital Communications Switching*, 1986 (Pitman) ISBN 0-273-02178-8

[17] Van Etten & Van Der Plaats, *Fundamentals of Optical Fibre Communications*, 1991 (Prentice Hall) ISBN 0-13-717521-3

[18] Senior, *Optical Fibre Communications*, 1985 (Prentice Hall) ISBN 0-13-638222-3

[19] Tugal & Tugal, *Data Transmission*, 1982 (McGraw-Hill) ISBN 0-07-065447-6

[20] Bartlett, *Cable Communications*, 1995 (McGraw-Hill) ISBN 0-07-005355-3

[21] Miller P., *TCP/IP Explained*, 1997 Digital Press ISBN 1-55558-166-8

[22] Miller S., *IPv6: The Next Generation Internet Protocol*, 1997 Digital Press ISBN 1-55558-188-9

[23] Trulove, *LAN Wiring*, 2000 McGraw-Hill ISBN 0-07-135776-9

[24] Manterfield, *Telecommunications Signalling*, 1999 IEE Press ISBN 0-85296-761-6

[25] Flood, *Telecommunications Networks*, 1997 IEE Press ISBN 0-85296-886-8

[26] Redl, Weber and Oliphant, *GSM and Personal Communications Handbook*, 1998 Artech House ISBN 0-89006-957-3

The following journals may also be useful for further reference.

[27] Compliance Engineering European Edition (Dash, Straus & Goodhue Inc.) ISSN 1080-2657

[28] Proceedings of the IEEE (The IEEE Inc.) ISSN 0018-9219

[29] Fiberoptic Product News (Gordon Publications)

[30] Electronics & Communication Engineering Journal (The IEE) ISSN 0954-0695

[31] British Telecommunications Engineering (The Institution of British Telecommunication Engineers) ISSN 0262-401X

Index

Introduction

The *Telecommunications Pocket Book* provides a summary of current telecommunications systems and working practice. Being a pocket book it cannot give full details of every system, but it will describe them in outline and act as an *aide mémoire*. Since telecommunication techniques are continually changing, the overview given will be useful to practising engineers who wish to update their knowledge.

It also contains data tables that will serve as a useful reference to technicians, engineers and students alike.

This book is structured to describe the signal sources first, followed by the equipment needed to convert these signals into an electrical format suitable for transmission, switching and processing. The transmission medium (radio, copper wire and optical fibre) is then described. Methods of modulating or digitizing the signals for transmission is described. Telephone exchange interfaces and switching is described before signalling and multiplexing are covered. Packet-based networks are described, including the Internet and the numerous protocols such as TCP/IP and HTTP that are associated with it. Standards and organizations are described last.

Telecommunication engineers use abbreviations extensively. This is unfortunate, but necessary to avoid very wordy descriptions of systems and techniques. Since this book will probably be read piecemeal, all abbreviations are described at the end of the book (Chapter 26). Useful formulae and data tables are included in this chapter.

The previous edition of the *Newnes Telecommunications Pocket Book* was published in 1998. Since then there have been many developments, particularly in mobile telecommunications. The growth of Internet traffic has spurned many other developments, such as Wavelength Division Multiplexing (WDM) over optical fibres, which are necessary to carry vast amounts of data. Another significant driver has been the de-regulation of telecommunications in many parts of the world. Regulations in many countries, including the US and most of Europe, have forced incumbent telecommunications companies to allow competition in the local loop (known as 'unbundling'). This, combined with Internet growth, has driven the development of digital line systems (xDSL) that allow high data rates into the home.

I would like to thank my family for their patience while I was editing this book at home during evenings and weekends. I would also like to thank fellow engineers at BT Laboratories for their interest and encouragement.

I have received feedback from readers, including university lecturers and engineers, and this has been welcomed. Please let the publisher know if you feel any topics need to be covered in more detail.

Steve Winder
2001

1 Telecommunications overview

Telecommunication engineering is about understanding the signals and the transmission medium, and then providing a means of communication in the most efficient way. The following sections describe speech, music, video and data signals. These are the signals we want to transmit from one place to another. Later chapters describe the means to transmit the signals (e.g. the telephone), the transmission path (e.g. optical fibre cable), switching and multiplexing. It is impossible to have dedicated circuits between every communication point, and it would be inefficient. Switching is used to connect points only for the duration of the call, which means that only the line between each communication point and the switch is dedicated – the rest of the telephone network is shared. Transmission paths between switch nodes are made to have a high capacity by multiplexing many circuits together. Instead of transmitting one telephone call along an optical fibre, we can transmit hundreds of calls.

Telecommunications encompasses a wide range of signal sources; it is truly multimedia. The telephone network has being integrated with computer systems to provide the Internet. Data is readily available from many sources and knowledge is king. We have become what has been called the 'information society'.

Historically, the telegraph arrived first; it was invented by Charles Wheatstone in 1837. Telegraph was made more efficient by Samuel Morse in 1844 by the development of the Morse code, which remains in use for some radio signalling. A telegraph code (using a 5-bit code to represent the alphabet) was introduced by Baudot in 1874. However, the typewriter was not invented until 1875, so the telegraph system was quite primitive until Western Union started to use the Baudot code in conjunction with a teletype (a sort of Telex machine) in 1901. The telephone was not invented until 1876 by Graham Alexander Bell, and switching was manual until the invention of the automatic telephone exchange by Strowger.

Many inventions have been developed since that time, the most significant was the computer. The computer has its beginnings during the war years 1939 to 1944, out of necessity to break the German navy's cipher codes. Commercial applications were developed and the US Space Program required small and light computers. The microprocessor and then the personal computer (PC) were developed. The computer explosion, where every company and almost every home has

2

a computer, has resulted in the need to share stored data. Computer data and video picture transmission are now common-place.

1.1 Speech and music signals

Speech requires a minimum of 300 Hz to 3.4 kHz bandwidth for good understanding. Speech is produced by a person when air from the lungs passes through the larynx (voice box) in the throat to cause a tone which is changed in amplitude by the mouth and tongue. Although the change in amplitude occurs at a low frequency, the tone is usually within the 300 Hz to 3.4 kHz bandwidth of a telephone system and no intelligibility is lost. Time varying speech signals and their frequency spectrum are illustrated in *Figure 1.1*.

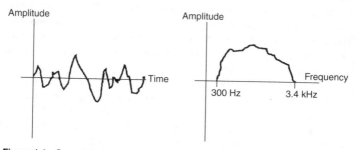

Figure 1.1   Speech in time and frequency domains

Music, on the other hand, requires up to 15 kHz bandwidth. Musical signals are similar to speech, and a passage of music can be identified if the bandwidth is restricted to the speech bandwidth of 300 Hz to 3.4 kHz. However, the quality of music is not appreciated in this limited bandwidth. Drum beats have a significant low frequency content and require a response down to a few hertz. Some string and wind instruments produce significant high frequency content sounds, some beyond the range of human hearing (which is up to 20 kHz in children, but this reduces with age and 15 kHz is a reasonable limit).

1.2 Video signals

Broadcast television signals are complex because they not only contain information about the brightness of the image, they also contain information about the colour and about picture synchronization. Sound signals are often transmitted with the picture too. The system works by the image in a camera being scanned left to right and top to bottom. On the display screen the picture is built up by brightening pixels (picture elements) by scanning them in the same order as the image in the camera. The camera and display scanning must be synchronized and the display refreshed 50–60 times per second. A colour television picture typically requires an analogue bandwidth of d.c. to 6 MHz. If digitized at a 16 MHz sampling rate and using 8-bit encoding, a data rate of 128 Mbit/s would be required.

Time varying video signals and the frequency spectrum for the PAL system is given in *Figure 1.2*. Other video systems such as NTSC or SECAM are similar. Each complete scan of an image generates a frame synchronizing pulse. This is followed by a burst of carrier at about 4.433 MHz to synchronize colour decoders in each receiver. The brightness of the image as it is scanned then determines the amplitude of the signal. Being an amplitude modulated signal, you might expect the spectrum to be double sideband, but it is not. The spectrum is vestigial sideband, which means that the lower sideband is filtered out beyond 1.75 MHz. This reduces the total bandwidth required, but still allows simple detection methods to be employed.

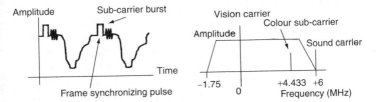

Figure 1.2 Video signals in time and frequency domains

Picture digitizing systems can reduce the bandwidth required. JPEG and MPEG encoding can reduce the transmission bandwidth of a television picture to 2 Mbit/s. Pre-recorded programmes can be digitized and stored. Decoding these signals takes place at the television in a set-top box. The encoding of 'live' television pictures is now commercially available.

Slow scan television is used for security surveillance systems. Generally these are monochrome pictures with no sound and are used to monitor slowly changing events, door entry systems, etc. By transmitting each picture slowly the bandwidth is much reduced. This system requires a picture store system at both the camera and the display terminal. A snap-shot is taken by the camera and the image

4

is digitized and stored. Data from the store is transmitted slowly over the transmission system and is stored at the display terminal.

Video-conferencing systems only transmit the difference between one image and the next. In the usual situation, where there is little movement, the amount of data needed to be sent is perhaps just 1% compared to the full picture (i.e. 1.28 Mbit/s). If the display is not refreshed often and video compression techniques are used, a bandwidth of only 64 kbit/s can be used.

1.3 Data signals

Computer data is a binary (two level) representation of numbers, text, images, etc. The binary number is either 1 or 0, often referred to as logic 1 and logic 0. A logic 1 is at or near the positive supply rail of the digital circuits. A logic 0 is at or near the 0V rail. The data word, or 'byte', is normally 8 bits long, but is broken down into two 'nibbles' so that the word can be expressed in hexadecimal form (0000 = 0, through to 1111 = F). Thus 7E = 01111110 in binary.

Data has a d.c. content. A string of logic 1s or logic 0s is clearly d.c. The maximum fundamental frequency occurs when the data string is 10101010, and the frequency is half the data rate. However, data is pulses which have significant harmonic content and a bandwidth of at least five times the data rate is required. Although data cannot be directly transmitted in a speech bandwidth of 300 Hz to 3.4 kHz, low speed data can be used to modulate a signal within that bandwidth. Techniques to remove the d.c. content of data do allow long range transmission. In one method, the data is converted into alternating positive and negative pulses to remove the d.c. content.

Figure 1.3 shows a data stream in both time and frequency domains. There is no spectral content in the signal at a frequency equal to the data rate, or at an integer multiplier of the data rate. If the data rate is 1 Mbit/s, so that each pulse is 1 µs long, the maximum fundamental frequency is 500 kHz. Since a square wave has no even

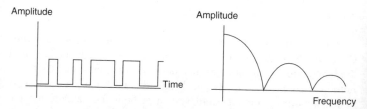

Figure 1.3   Data signal in time and frequency domains

harmonic content, there is no spectral power at 1 or 2 MHz. There are, however, peaks in the spectral content at 1.5 and 2.5 MHz.

Data rates for transmission are expressed as either bits per second or bytes per second. The higher the data rate, the lower the time required for transmission and the processor can begin another task. If a transmission link is charged on a per second basis, the higher data rate also means less cost. For both of these reasons, modem speeds have increased as the technology has permitted and are close to their theoretical limit.

Another reason for increased data rates is the use of interactive systems. Automatic teller machines which dispense cash through a 'hole in the wall' are an example of interactive systems, however the data rate can be quite low without seriously affecting performance. Where data rate is very important is in Internet applications. The Internet response time is notoriously slow, with some pages taking a few minutes to download.

2 Telephone equipment

2.1 Constraints

This chapter deals with the terminal equipment that can be connected to the local distribution cables of a public switched telephone network (PSTN), and therefore covers equipment that can be installed at domestic and small business premises, or extensions on a PBX using local lines.

Public Telephone Operator (PTO) monopolies are being broken worldwide to enable competition from privately run companies. This, and the rapid advance of integrated circuits, has resulted in a wide choice of both facilities and suppliers. This is the lowest level at which connections can be made to the PSTN, and imposes a number of constraints on the equipment:

(a) The available frequency band is 300 to 3400 Hz. This constraint is imposed by the 8 kHz sampling rate of analogue to digital converters and the limitations of the transmission bridges used in the local exchanges. A DC path exists for signalling purposes but not transmissions.

(b) The −50 volt exchange battery supply is fed out to the local lines. This is provided primarily to give a DC loop calling facility, and to energize a carbon microphone. It does, however, provide a small amount of available power for the operation of terminal equipment, and with low-power ICs enables a considerable amount of processing to take place in the terminal equipment without the provision of local power. The available power is limited by the line loop resistance of up to 1250 ohms. Constant current feeds are supplied from some digital exchanges.

(c) It is an offence to connect unapproved equipment to the public network. The requirements of telephone apparatus to meet certain standards depends on the local telecommunications regulator. Generally equipment has to meet the EMC and safety standards of the host country. The equipment must meet specified requirements for that class of equipment, and be tested to the appropriate specifications by an authorized independent body before approval is granted. If the equipment derives power from the mains supply, stringent safety conditions apply.

2.2 Telephones

In its simplest form a telephone consists of:

(a) a microphone and earpiece, normally combined into a single hand-held unit;
(b) means of signalling the exchange when a call is to be made;
(c) means of sending the 'address' information of the called subscriber to the exchange;
(d) an incoming call alarm;
(e) means of signalling the exchange that the call has been answered.

The requirements of (b) and (e) are met by completing a DC loop when the handset is removed from its rest. The DC path is then via a carbon transmitter, providing the necessary polarizing current for the transmitter. Carbon microphones developed for the telephone provide an adequately good frequency response for speech at high sensitivity, avoiding the need for amplification. Less sensitive microphones with amplifiers are now more common having longer life and avoiding the problem of granules breaking down and causing noise.

The incoming call alarm (d) has in the past been provided by a bell operating from a 17 Hz supply from the local exchange. This is being largely replaced by an audible tone from a small sounder or loudspeaker, a choice of tone being available in some equipment to allow a number of phones in a restricted area to be identified. This is referred to as a 'tone caller' and on some equipment the earpiece is used for this purpose. In such cases the calling signal must start at a low level and progressively increase as a safety precaution.

The 'address' of any subscriber is represented by a series of digits, the first group in the series providing the routing information to the local exchange, and the second group, the subscriber's individual number in the exchange. Since the advent of automatic exchanges the series of digits has been provided by a dial which produces an impulse sequence, breaking the DC path. The impulses in the train are $33 \left(\frac{1}{3} \right)$ ms on, $66 \left(\frac{2}{3} \right)$ ms off, providing an impulsing rate of 10 per second. The speed restriction was imposed by the switching rate of the mechanical selectors and not by the transmission path; hence the introduction of electronic switching has made possible a much faster switching rate, and reduced the time required to set up a call.

Press-button switches have therefore replaced dials on modern telephones. When used in areas still served by mechanically switched exchanges, the keypad is quicker and more convenient to use, but has store the number and send it to line as a 10 ips signal; hence the actual time required to set up the call is not substantially reduced. When operated into a digital exchange, however, multifrequency coding is employed, each digit being represented by two audio tones. DTMF signalling allows 10 digits, plus * and # symbols to be coded as shown in *Table 2.1*.

Table 2.1 DTMF code

Digit	Frequencies (Hz)	
1	697	1209
2	697	1336
3	697	1477
4	770	1209
5	770	1336
6	770	1477
7	852	1209
8	852	1336
9	852	1477
0	941	1336
*	941	1209
#	941	1477

Many of the telephones now in production will produce either type of signal with a simple changeover to enable them to operate with either type of exchange.

The logical extension to the keypad is number storage, and a wide range of telephones incorporate both last number redial and a store enabling the most used numbers to be called by depressing a single press button. Again the maximum advantage is obtained from digital exchanges, since exchanges requiring impulse dialling cannot take advantage of the quick-call set-up obtained with MF coding. Further facilities now available include an LCD display, which shows the number being called, and can also provide a clock and call timer, and a loudspeaker to allow handsfree operation.

The digital exchange is also increasing the services available to the customer. These include:

Charge advice: enabling the cost of a call to be advised at the end of the call.
Reminder call: enabling an alarm call to be made at any required time.
Call diversion: enabling all incoming calls to be diverted to another telephone.
Call barring: enabling certain types of call (e.g. overseas or trunk) to be barred.

9

Call waiting: provision of a tone to indicate that another call is waiting to be connected.

Three-way calling: a facility enabling a third subscriber to be dialled during a call, and included in the conversation.

Freephone: a service involving no charge to the caller on numbers where the costs are met by the called party.

Full duplex operation, that is the ability to converse in both directions, is normally required from a telephone. This is simple with a four-wire system where go and return are separate circuits, but connection to the PSTN requires two-way working. Superposition of the transmitted and received signals results in a caller hearing his or her own voice in the receiver at an unacceptably high level. This is known as sidetone, and the level at which this occurs has a significant effect on talk volume. Complete absence of sidetone (as occurs with a four-wire system) provides a 'dead' impression and the caller tends to raise his or her voice. Too high a level is disturbing to the caller, causing the caller to speak more quietly.

Telephones therefore contain the means for cancelling part of the sidetone, a facility normally provided by a hybrid transformer. *Figure 2.1* shows a typical telephone circuit (dialling and switching omitted) incorporating an anti-sidetone induction coil, and a chain of shunt diodes which limit the level transmitted to line when the line loop resistance is low.

2.3 Loudspeakers

Moving-coil loudspeakers in which the 'speech' coil is suspended in the gap of a powerful permanent magnet are universally used. The coil is centred in the gap by a fibre or plastic 'spider' and attached to the small end of a paper or fibre cone whose wide end is attached to the speaker frame by a flexible surround. Audio-frequency currents flowing in the coil produce magnetic fields which react with the permanent magnet to produce motion of the coil and cone. The sound pressure waves generated by cone movement reproduce the audio signal in the coil.

In normal telecommunications high fidelity is not a criterion as it is in a music system and only reasonable quality of speech reproduction is required, necessitating a bandwidth of, say, 300–3500 Hz. This can satisfactorily be achieved with quite small-diameter speakers (i.e. cm) and without specially designed enclosures since extended bass response down to 50 Hz or less is unnecessary. The efficiency of conversion of electrical to acoustic energy is quite low – in the order 5–20%.

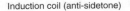
Induction coil (anti-sidetone)

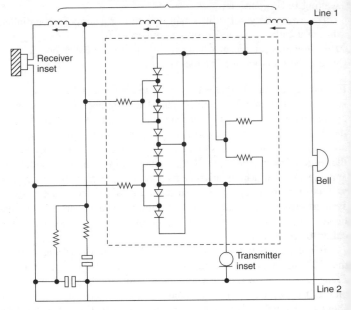

Figure 2.1 Telephone circuit

Loudspeakers are specified by power and impedance; the quote
power is an electrical rating and may not be a true RMS figure a
sometimes an artificial 'RMS music' power or 'peak rating' is quote
These are based on the practical consideration that the actual sign
applied to the speech coil is non-sinusoidal. The impedance quote
for a loudspeaker (which has both resistance and inductance) is th
lowest value measured above the frequency of mechanical resonanc
Typical values are 4, 8, 12 and 16 ohms although some miniatu
units may be as high as 64 ohms.

Loudspeakers are normally voltage driven from a nominally ze
impedance source (achieved by shunt negative feedback) but instanc
may be found where a transformer is used to match the speaker no
inal impedance to a higher impedance source.

2.4 Microphones

Microphones in current use may be carbon granule, moving coil, va
able capacitance, piezo-electric or based on the 'electret' which